Tipbook
Amplifiers
The Complete Guide and Effects

Hugo Pinksterboer

Tipbook
Amplifiers
The Complete Guide and Effects

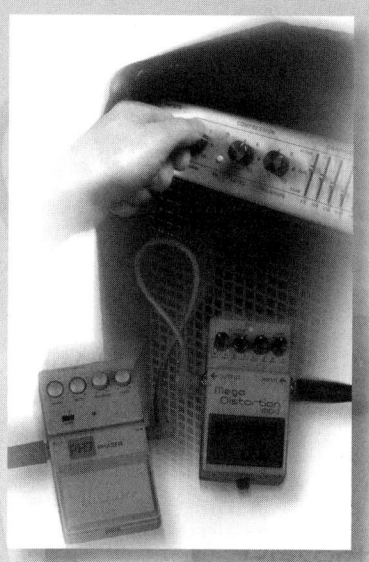

HAL•LEONARD®

The Complete Guide to Your Instrument!

First edition published in 2005 by The Tipbook Company bv,
The Netherlands
Second edition published in 2009 by Hal Leonard Books
An Imprint of Hal Leonard Corporation
7777 West Bluemound Road
Milwaukee, WI 53213

Trade Book Division Editorial Offices
19 West 21st Street, New York, NY 10010

Printed in the United States

Book design by Gijs Bierenbroodspot

Library of Congress Cataloging-in-Publication Data

Pinksterboer, Hugo.
   Tipbook amplifiers and effects : the complete guide / by
   Hugo Pinksterboer.
        p. cm.
   Includes bibliographical references and index.
   ISBN 978-1-4234-6277-4 (alk. paper)
 1. Electronic musical instruments. 2. Audio amplifiers. I. Title.
   ML1092.P57 2009
   786.7'19—dc22
                              2009005097

www.halleonard.com

Thanks!

For their information, expertise, time, and help, we'd like to thank the following musicians, technicians, and other experts:

J. Hayes (PRS), Stephen White (Guitar Tech, CA) Steve Fidler (Hiwatt, UK), Elliot Freedman, Dolf Koch (Koch Guitar Electronics), Bryan Beller (SWR), Richard Fleming (Laney/BLT Industries), Doug Reynolds (US Music Corporation), Jan Betten (SD Systems), Koos Hofman, Harry de Jonge, Paul de Jong, Jos Kamphuis, Ernst Fliek (Ernst Amps), Ferry Verhoeve, Rob de Vos (Hevos bass amps), Mark Zandveld, Tjako Fennema, Luc Wäckerlin, Remco Drubbel, Tim Benniks, Edwin in 't Veld (Heuff Sound and Vision), Bert Smorenburg, Nicky Moeken, Rempe Kooij, Jan Tonnis, Pim Dollé, Christian Robbemond, Edwin Kool (USA Music import), Fred Kienhuis, Henny van Ochten (Texas & Tweed), Allard Krijger (Interface), Alfons Verreijt, Eric Rutten (TM Audio/Shure), Mr. Lindner (Sennheiser), and Arie-Jan Folkerts (Marble/FNS Tube Technologies).

About the Author

Journalist, writer, and musician **Hugo Pinksterboer**, author of *The Tipbook Series*, has published hundreds of interviews, articles, and reviews for national and international music magazines.

About the Designer

Illustrator, designer, and musician **Gijs Bierenbroodspot** has worked as an art director for a wide variety of magazines and has developed numerous ad campaigns. While searching in vain for information about saxophone mouthpieces, he got the idea for this series of books on music and musical instruments. He is responsible of the layout and illustrations for all of the Tipbooks.

Acknowledgments

Cover photo: René Vervloet and Gijs Bierenbroodspot
Editors: Robert L. Doerschuk and Meg Clark
Proofreaders: Matt Blackett

Anything missing?

Any omissions? Any areas that could be improved? Please go to www.tipbook.com to contact us, or send an email to info@tipbook.com. Thanks!

Contents

X Introduction

. . . .

XII See and Hear What You Read with Tipcodes
www.tipbook.com
The Tipcodes in this book give you access
to short videos, sound files, and other
additional information at www.tipbook.com.

. . . .

1 Chapter 1. Amps and Effects
What they are and what they do.

5 Chapter 2. A Quick Tour
An overview of the main types of amps and an
introduction to effects.

27 Chapter 3. Buying Equipment
What amps and effects cost, what you're paying for,
and where to buy new or used gear.

35 Chapter 4. Features
A quick run through of the basic features of amps
and effects. Includes information on speakers, digital
equipment, and MIDI.

77 Chapter 5. Figures
A practical, non-technical introduction to the various
figures you need to deal with when selecting or using
amps and effects.

91 Chapter 6. Testing Amps: General Tips
General tips on testing amps and effects. Specific tips
for various amp types are covered in chapters 7 to 11.

VII

97   **Chapter 7. Guitar Amps**
Features and testing tips.

123   **Chapter 8. Bass Amps**
Features and testing tips. Includes tips on amplifying upright basses.

137   **Chapter 9. Acoustic Amps**
Features and testing tips.

145   **Chapter 10. Keyboard Amps**
Features and testing tips.

151   **Chapter 11. Sound Reinforcement / PA Systems**
Features and testing tips.

171   **Chapter 12. Effects**
The various types of effects, what they do, and what to listen for. Includes information on connecting and testing effects.

199   **Chapter 13. Effect Equipment**
Effects come in various guises, from built-in effects to stompboxes and rack-mounted multi-effects units.

215   **Chapter 14. Microphones and Pickups**
Basic information on dynamic and condenser microphones and their patterns, the various types of instrument pickups, pickup/mic combinations, and more.

229   **Chapter 15. Cables and Wireless Systems**
The differences between microphone, instrument, and speaker cables, what makes a good cable, and the basics on wireless systems.

241   **Chapter 16. Care and Maintenance**
Practical tips for at home and on the road.

VIII

251 **Chapter 17. History**
A brief history of instrument amplification and effects.

255 **Chapter 18. Brands**
An introduction to the many brand names.

261 **Glossary**
What is bi-amping, what does a DI box do, and what does TRS stand for? A handy glossary for anyone who uses amps and effects.

272 **Tipcode List**
All amp and effect Tipcodes listed.

273 **Want to Know More?**
Information about magazines, websites, and other resources.

277 **Essential Data**
Three pages for essential notes on your equipment and favorite settings.

282 **Index**
A complete index of terms.

289 **The Tipbook Series**
Brief descriptions of the other volumes in the Tipbook Series.

Introduction

Are you planning to buy a new amplifier? Would you like to add new effects to your current rig? Do you want to learn more about the equipment you're using? This book will tell you all you need to know, no matter which instrument or which style of music you play — from power ratings to preamps and pickups, from cables to cabinets, from microphones to monitors, from tubes to tweeters, and much, much more. Loaded with practical, musical tips and advice, Tipbook Amplifiers And Effects covers even the most technical subjects in a non-technical way.

After reading this book, you'll be able to get the most out of your equipment, to buy the gear that best suits your needs, your preferences, and your budget, and to easily grasp most literature on the subject, from catalogs, manuals, and magazines to books and online publications.

The first three
If you're a first time buyer, please check out the first three chapters. They will introduce you to the different types of amps and effects, and explain the basic terminology and features of this type of gear. Basic price indications and buying tips are included in chapter 3.

Features
Chapter 4 tells you everything about the main parts, the controls and the features that you'll find on this type of equipment,

from knobs and faders to equalizers, balanced and unbalanced inputs, DI, speaker simulation, effects loops, power conditioners, and numerous other subjects —all written for musicians, not technicians.

The figures

Dealing with amps is dealing with figures. You need to know a little about power ratings and decibels, and it comes in very handy if you know the basics about frequencies and Ohms. Chapter 5 reveals how accessible this type of information can be.

Amps

The next four chapters each deal with a certain type of amp, explaining their features in more detail, followed by a chapter on sound systems.

Effects

The intriguing subject of effects is covered in two separate chapters. The first one explains the different types of effects, from reverbs and wahs to jackhammers and ice boxes. The second chapter explains effect equipment in detail.

Microphones, cables and more

This book wouldn't be complete without special chapters on microphones and pickups, on cables and wireless systems, and maintenance. As an extra, there's a brief chapter on the history of live amplification, and one that introduces you to the main companies that produce amps, effects, microphones, and speakers.

Glossary and index

The Tipbook glossary briefly explains most of the terminology used in the previous chapters, and the index makes the information in the book really accessible. Enjoy!

— Hugo Pinksterboer

See and Hear What You Read with Tipcodes

www.tipbook.com

In addition to the many illustrations on the following pages, Tipbooks offer you another way to see—and even hear— what you are reading about. The Tipcodes that you will come across regularly in this book give you access to extra pictures, short videos, sound files, and other additional information at www.tipbook.com.

Here's how it works. On page 73 of this book, there's a paragraph on how important speakers are to your sound. Below that paragraph is a short section marked **Tipcode AMPS-007**. Type in that code on the Tipcode page at www.tipbook.com and you will hear a brief demo of the sounds of a number of different speakers. Similar audio examples are available on a variety of subjects; other Tipcodes will link to a short video.

TIPCODE

Tipcode AMPS-007
Play this Tipcode to hear the sounds of five different speakers in the same combo amp.

XII

Repeat

If you miss something the first time, you can of course replay the Tipcode. And if it all happens too fast, use the pause button below the movie window.

Tipcode list

For your convenience, the Tipcodes presented in this book are listed on page 272. The Tipcodes in this book include demonstrations of different amp models and a wide variety of effects.

Plug-ins

If the software you need to view the videos is not yet installed on your computer, you'll automatically be told which software you

First, make your selection: Tipcode, chords and fingering charts, or the glossary.

The Tipcode window displays videos, fingering charts, chords, or a glossary of the terms used in this book.

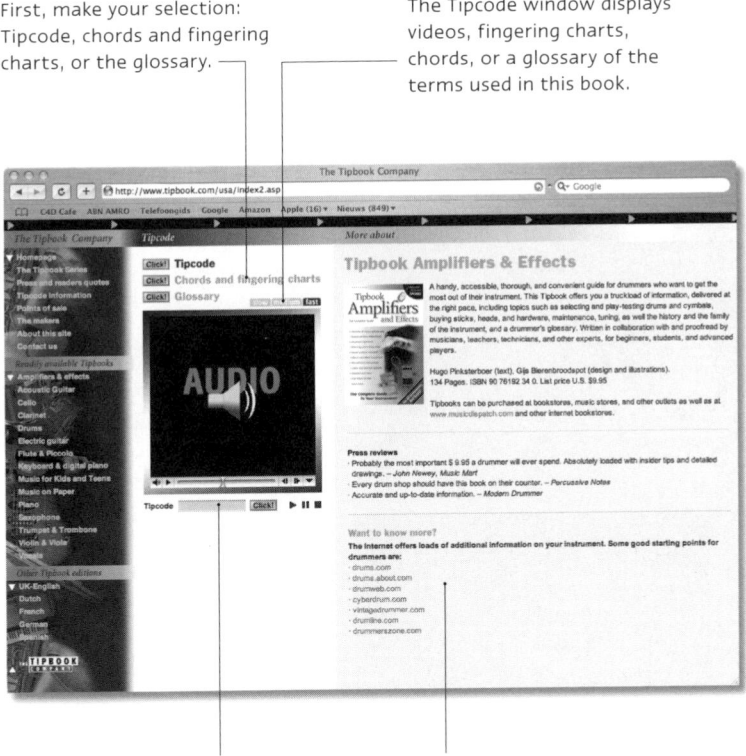

Enter a Tipcode here and click on the button. Want to see it again? Click again.

These links take you directly to other interesting sites.

XIII

need, and where you can download it. This type of software is free. Questions? Check out 'About this site' at www.tipbook.com.

Still more at www.tipbook.com

You can find even more information at www.tipbook.com. For instance, you can look up words in the glossaries of all the Tipbooks published to date. There are chord diagrams for guitarists and pianists; fingering charts for saxophonists, clarinetists, and flutists; and rudiments for drummers. Also included are links to most of the websites mentioned in the Want to Know More? section of each Tipbook.

Tipbook
Amplifiers
and Effects

1

Amps and Effects

Without amplifiers, the world of music would be a different place. There would be no electric guitars, basses or synthesizers. No rock 'n' roll. No metal. No big stadium shows, and so on.

Of course, without amplifiers you wouldn't be able to enjoy music at home or on the road either. The type of amplifier in your home stereo, your MD player or your car, however, is quite different from the amps that are used on stage — and the same goes for their speakers.

Types

There are many types of instrument amplifiers and speakers, each designed with a specific purpose. Bass guitars require powerful amps and special speakers to faithfully reproduce the instrument's low notes. Keyboard amplifiers need to deliver the entire tonal range of synthesizers and digital pianos naturally and without distortion, from the lowest to the highest notes — and the same amps can be used for vocals and acoustic instruments, too. For electric guitarists, the amplifier is as important to their sound as their guitar. Unlike most other types of amps, guitar amplifiers are designed to strongly influence and color the sound of the instrument.

Combo amp

The instruments mentioned above don't do you much good without proper amplification. Home keyboards and digital pianos usually have a built-in amp and speakers, but almost all other electric and electronic instruments require a dedicated amp. Often, this will be a unit that has both the amplifier and one or more speakers in a single box: a *combo amp*. Many combo amps feature one or more built-in effects as well.

Acoustic instruments

Saxophones, drums, pianos, flutes, violins, and thousands of other instruments can be played without amplification: They are *acoustic* instruments. Amplification is required only if you play them in a band that features instruments that sound a little or a lot louder, or if you play for more than just a few people.

The bassist first

In an acoustic jazz group (e.g., piano, upright bass, drums, and a saxophone), the bassist is the first who may need an amp: The double bass or upright bass doesn't produce a whole lot of sound.

2

If bass is not amplified, the drummer will need to play extremely softly. If the piano is amplified, some sound reinforcement might be required for the horns as well. The drummer can still be heard, unless the band plays larger venues.

Electric
Drums can be played loud enough to be heard in a rock band, as long as you play small rooms only. In bigger places or in louder bands, you won't be able to hear the bass drum anymore. As the volume level goes up, the rest of the drum set might get lost in the din as well.

The entire band
When the entire band needs to be amplified, you'll need a *sound reinforcement* or *PA (public address) system*. Chapter 11 of this book provides basic information on sound systems.

Microphones
Microphones are used by vocalists, as well as by horn players, drummers, and others who play acoustic instruments. They're also used in front of guitar and bass amps, transmitting the sound from their speakers to the sound system, where it can be balanced with the sound of the other instruments. Guitarists, violinists, bassists and other musicians often use *pickups* instead of microphones. Both microphones and pickups are briefly dealt with in Chapter 14.

Effects
Effects relate to amplifiers as spices do to food: You can do without them, but they make things a whole lot tastier and they add a whole lot of variations. From ambient reverbs and thick, liquid choruses to effects that make you sound like you're from outer space — they're all covered and explained in Chapters 12 and 13.

3

2

A Quick Tour

This chapter introduces you to the main parts and controls of a basic amplifier. It tells you the differences between the main amplifier types, and it covers the basics of what effects are and what they're used for.

When you play a CD on your home stereo, the CD player sends the signal it picks up from the CD to the amplifier. The amplifier boosts the signal and sends it to the speakers, so you can hear it. Instrument amplification works basically the same.

Controls

If you look at the control panel of a guitar or a bass amp, you'll see some familiar terms. First, there is a volume or *master volume* control. This sets the overall volume level of the amplifier.

Tone control

Next, there are one or more *tone* controls. Usually, you can adjust the amount of *bass* and *treble* separately, just like on your home stereo. Turning the bass knob up makes for a bassier, warmer, fatter sound. Adjusting the treble control affects the high range, making the sound either duller (at low settings) or brighter (at higher settings). Treble and bass controls sometimes sport different names such as *hi-trim* and *lo-trim control* or *bottom* and *edge*, respectively.

Mid

Instrument amplifiers often have a third knob, which cuts or boosts the *midrange* of the sound spectrum. If you want a singer to sound more intelligible, you'll probably have to boost this range, while guitarists who go for a heavy crunch sound will turn this control down.

Equalizer

The tone-shaping controls are collectively known as the *equalizer* or *EQ*. An equalizer with separate controls for bass, midrange and treble is called a *three-band* equalizer.

TIP

More bands and controls

Some amps have extended tone control options: an equalizer with more bands, for example, or switches that make the entire sound brighter or punchier, or a presence control that makes the sound more 'present' by boosting the uppermost treble frequencies.

6

The room and your instrument

Equalizers can be used to adapt your sound to the room: In a dull sounding room, you turn up the treble knob a little, nudging the higher frequencies. However, you can also use these tone controls to color and shape the sound of your instrument.

instrument input (jack)

headphone output (jack)

grille   volume bass   treble

power switch

A very basic combo amplifier.

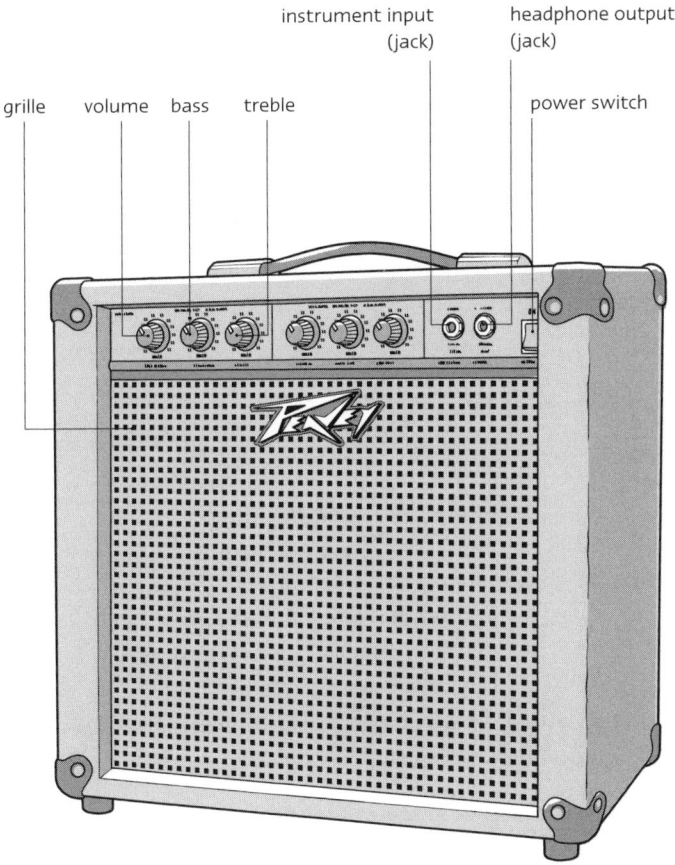

Gain

The *gain control* sets how much signal goes into the amp. If you turn it up too far, the sound will distort. Most musicians try to avoid this effect — but most guitarists love it, and so do others, including bassists, harmonica players (see page 18), and keyboardists.

7

TIPCODE

Tipcode AMPS-001
Opening up the gain control of a
guitar amp produces a this
— desired—type of distortion.

INPUTS AND OUTPUTS

You connect your instrument to the amp's *instrument input*, which
is usually located on the front side or the top of the combo amp.
Microphones have special inputs.

More instruments
Many amps have several inputs, allowing you to connect two or
more instruments, an instrument and a microphone, or a stereo
instrument such as a piano or a keyboard.

Regular or hot
Guitar and bass amps often have two or more instrument inputs
for a different reason: One is for instruments that produce
a regular signal, and the other is a less sensitive input for
instruments with a 'hotter,' stronger signal. Some amps have a
separate *bright input*, for a brighter, more trebly sound.

TIP

Two instruments
While not intended for that purpose, you can often use those
two inputs to connect two instruments simultaneously (see
pages 109–110 and 127).

Headphones
Just like home stereo amps, many instrument amplifiers have a

8

headphone jack so you can practice without bothering others. *Tip:* connecting a headphone may or may not automatically disconnect the built-in speaker! Always check if it's really off when you're using headphones.

More ins and outs

Most amplifiers have a number of additional *input* and *output jacks*, usually located at the back panel of the amp. Here are some examples:

- An **auxiliary input** allows you to connect a CD player, a drum machine, or a keyboard, for example.

- If the amp has an **extension speaker output**, you can connect a separate *speaker cabinet* or *cab*.

- A **line out or recording out** can be used to hook the amp up to a PA system, a recording device, or the input of your computer's sound card, for example.

- **Send and return** or **insert** sockets allow for the connection of an effects unit.

- A **footswitch** input allows you to control built-in effects or select channels (see page 11) with your foot.

In or out?

The terms 'input' and 'output' can be confusing — and the fact that their jacks often look identical doesn't help. But it's not that complex, really, as long as you remember that the signal always flows from one output to another input.

- Your instrument has an **output jack**: When you connect a cable, this is where the signal goes *out* of the instrument, toward the amp.

- At the amp, the instrument cable is connected to the instrument **input**, where the signal goes *into* the amp.

- From the amp, the signal may go to an external speaker: It will leave the amp at a **speaker output**. From there it travels to the *input* of the external speaker cabinet.

9

- In a nutshell, signals always go from one output to another input. **Never** try to connect inputs to inputs, or outputs to outputs!

Jacks are also known as *sockets* or *receptacles*.

SPEAKERS

Electric guitar combos usually have a single speaker, or two or four identical ones. Combos with two or more speakers typically cost more, of course.

Two-way and three-way systems
Most bass and keyboard combos have two different speakers or *drivers*: a relatively large *woofer* for the lower sound range, and a much smaller *tweeter* for the high frequencies. This is known as a *two-way system*. Some units have a separate *midrange* driver too. Then you're talking about a *three-way* system.

Speaker sizes
Most combos feature a 10", 12", or even a 15" woofer. Practice amps and special amps for acoustic guitars often use smaller speakers. Tweeters can be a small as 1" (ca. 2.5 cm). They are often mounted on a *horn* that helps disperse the high frequencies. Electric guitar combos don't have tweeters as they make a distorted guitar sound like a bee in a box. Larger speakers (e.g., 18") are mainly used in dedicated bass cabinets.

The same
Some combos and cabinets feature a number of identical speakers. Bass combos and cabs, for example, may feature four or eight small 8" or 10" speakers, plus a tweeter. Together, these small speakers are capable of reproducing the low frequencies that are essential in bass amplification.

Sound system
The speaker cabinets of a small sound system often use 12" or 15"

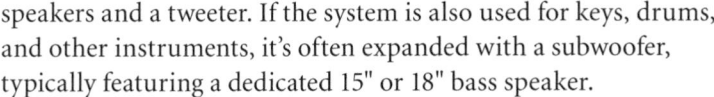

speakers and a tweeter. If the system is also used for keys, drums, and other instruments, it's often expanded with a subwoofer, typically featuring a dedicated 15" or 18" bass speaker.

Baffle board

The speakers are mounted on a *baffle board*.

THE AMPLIFIER

Amplifiers actually consist of two main sections. The first section is the *preamplifier* or *preamp*. This is the section where you plug your instrument or microphone in. The second section is the *power amp*.

Preamp

The preamp's controls allow you to shape your sound and control your overall volume: you use its master volume control, the tone controls, the gain control and all other controls to adjust your sound and the overall volume.

Power amp

The power amp is the workhorse. It boosts the signal that comes from the preamp, and sends it to the speaker(s).

Integrated amplifier

In most cases, the preamp and the power amp share the same housing. This is known as an **integrated amplifier**.

TIP

Channels

Amplifiers often have two or more *channels*. With two channels it's almost like having two amplifiers. If every channel has its own input, you can connect a different instrument to each one. This is common with keyboard amps, for example.

11

Guitar amps

Guitar amps often have two or more channels too, but these channels are typically used to create different types of sound for a single guitar. On a two-channel guitar amp, the first channel usually provides a clean sound, while the other generates an overdriven, distorted sound.

Shared or independent

Each channel may have its own *independent* tone controls, but they might be shared as well. The same goes for all other types of controls.

POWER

The actual power of the power amp is commonly expressed in *watts RMS*. If you want to compare the *power ratings* of a number of amplifiers, this is type of figure you should look for — but do note that power ratings don't tell the whole story about how loud amps are, as you will read on page 78.

Speaker watts

The amount of power a speaker can handle is also expressed in watts. This figure doesn't say a thing about how much sound a speaker can produce, as you will see on page 80.

How much

So how much power do you need?

- If you're a guitarist, a 10 to 15 watt amp will usually do just fine for **practicing at home**, and the same goes for most other musicians.

- For **rehearsals and small club gigs**, guitarists often use a 30 to 60 watt amp. If you use a tube amp (see pages 101–102), 10 or 15 watt may be plenty.

- As low notes require more power to be properly amplified, the **bassist** in that same band will commonly use 100 watts or more.

12

- Around 50 to 75 watts will allow the **keyboard player** to keep up.

- Small **sound systems** start around 75 to 150 watts per channel. (They're stereo, so they have separate right and left channels, just like your home stereo.)

SMALL AND PORTABLE

The very smallest amps are headphone amps for bass or guitar. Some are so small that you can actually plug the entire amp in your instrument; no cables required. They typically feature a limited number of sounds, and controls for gain, tone, and volume. Slightly larger pocket-size amplifiers may provide you with numerous features including a variety of guitar tones and effects, amp models, drum and bass grooves, demo songs, a tuner, a phrase trainer (see page 212), and an input for a sound source that allows you to jam along with your favorite recordings.

Battery-powered amps
If you want to play outside, there's a limited number of battery-powered combo amps available with power ratings up to 30 watts. Next to a guitar input, they often feature inputs for other instruments and a microphone.

STACKS

Rather than a combo, guitarists and bassists often use a separate amplifier (*amp head*, *head* or *top*) combined with one or more *speaker cabinets*, *cabs* or *enclosures*. These cabs typically feature two or more speakers.

Full and half stacks
A *full stack* consists of a head and two cabs; a *half stack* has one

13

cab. Stacks are typically easier on the road than equally powerful combo amplifiers; that said, most of these *piggyback* combinations are way too heavy to be built into a single housing. Using a stack also allows you to choose your own amp/speaker combination, and it often looks more impressive than a combo amp, too.

4x12, 2x10
Speaker cabinets or cabs typically have two or more speakers. Many guitarists use a cabinet with four 12" speakers, indicated as *a 4x12* or *412*. Bassists often use 2x10 (210) or 4x10 (410) cabs, with two and four 10" speakers respectively. Others prefer an 810 cab, for example.

A half stack.

head or
amp head

cab

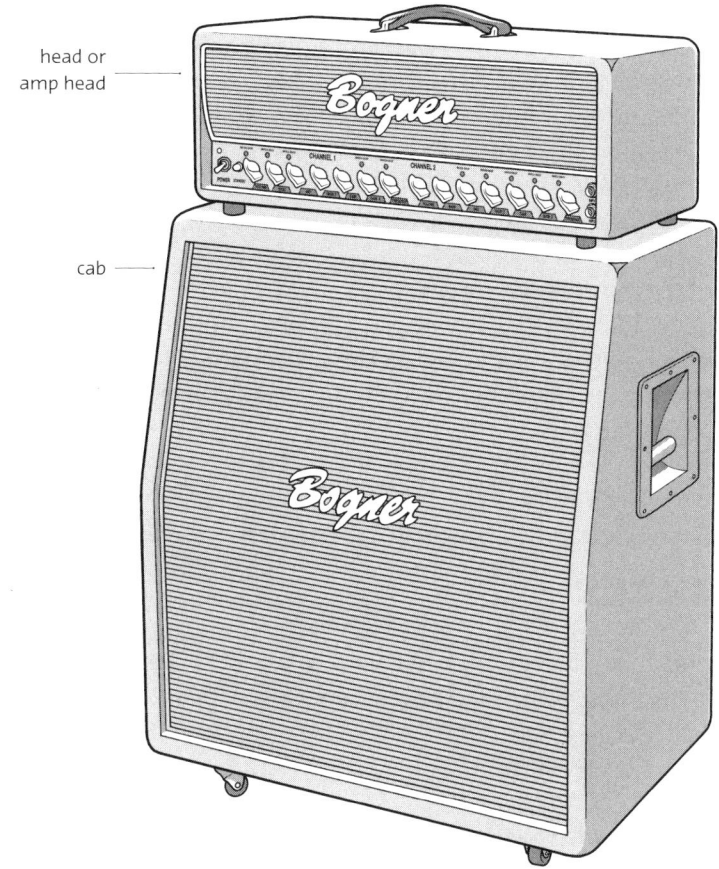

RACKS

A *rack system* consists of a frame that can hold a number of *rack-mountable* devices: a preamp, a power amp, effects units, and so on. Rack-mountable equipment has a standard width of 19". Their height is expressed in the number of *rack units* (U or RU) or *rack spaces* they occupy. A *2U* or *2*-space device takes up two rack units.

Maximum flexibility

Racks systems offer maximum flexibility. You can choose every component yourself, which requires both experience and knowledge. Of course, rack systems are usually built into a case.

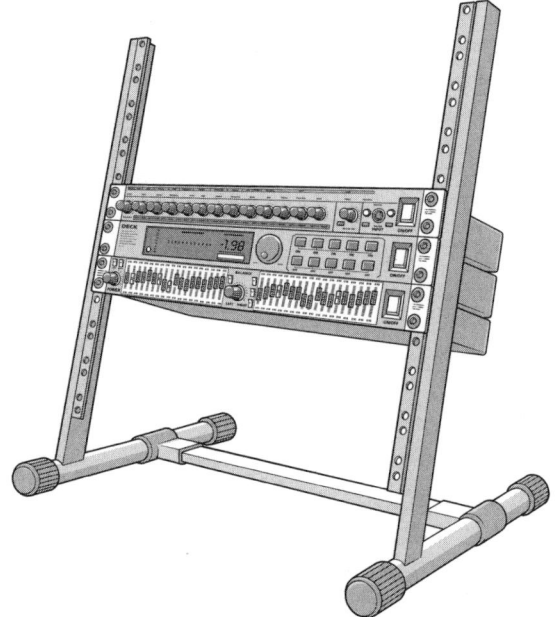

A rack system.

SOFTWARE

You can turn your computer into a home studio amp simulator. Various companies sell *software amps*. These virtual amps provide the same types of controls and features as the amps covered in this book, often with far greater flexibility. Some of these programs are stand-alone applications; others are plug-ins, which need a so-called host (i.e., music production software such as Cubase, Pro Tools, or Logic), or they can be used both stand-alone and as a plug-in.

Studio and live

Originally, this type of software was designed for (home) studio use only, but later editions and releases can be used live as well, completed by a *foot controller* that you plug your instrument into. This controller connects to your computer's sound card or USB port.

TIP

Connecting your guitar

A growing number of guitars come with a USB-port, allowing you to directly hook up your instrument to your computer. The software that comes with the guitar offers access to numerous guitar sounds, and it typically allows to record your music as well.

Regular guitars and basses need an audio interface. Interfaces can be built into a special cable (featuring a ¹/₄" jack for the guitar on one side, and a USB connector on the other), or they're separate units. Extended interfaces may feature built-in speakers, as well as a music player, a microphone input, and CDs with backing tracks to play along with.

TYPES OF AMPLIFIERS

There are special amps for electric guitar, bass, and keyboards, but also for acoustic/electric guitars and even for harmonicas and electronic drums. The latter two are quite rare. Though instruments usually sound best when used with a dedicated

16

amplifier, you can use your electric guitar amp for your acoustic guitar, too. Some keyboardists and acoustic guitarists love the sound of certain bass amps, there are bass amp designs that became extremely popular among electric guitarists, and electric violinists sometimes use electric guitar amps.

GUITAR AMPS

Most amps are supposed to do one main thing: amplify. Guitar amps do more: They color the sound. A guitar amp really is half of the guitarist's instrument and represents half of the tone, rather than a device that simply makes the guitar sound loud.

Back in time
When pop music started to emerge, the need for more volume soon followed. Guitarists simply cranked up the low-powered amps of that time, making the sound distort — and it just so happened that this 'undesirable' side effect became a standard in guitar sounds.

Gain
Today's guitar amps often have a separate channel for this effect. It's usually labeled *drive, overdrive, lead, solo,* or *crunch channel.* The more you crank this channel's gain control, the more you will make the preamp distort. This can produce anything from a subtle edge to an intense, brutal, fully distorted tone. The sound will also get louder, which you can compensate for by turning the *master volume* control down.

Channel switching
The other channel is the *clean channel.* Switching from channel to channel — from a clean sound to a distorted tone -— is simply a matter of pushing a button that's located at the amp's front panel or on an external footswitch, or both. Some amps feature three or four channels, offering direct access to three or four guitar sounds, typically with increasing degrees of distortion (see page 108).

17

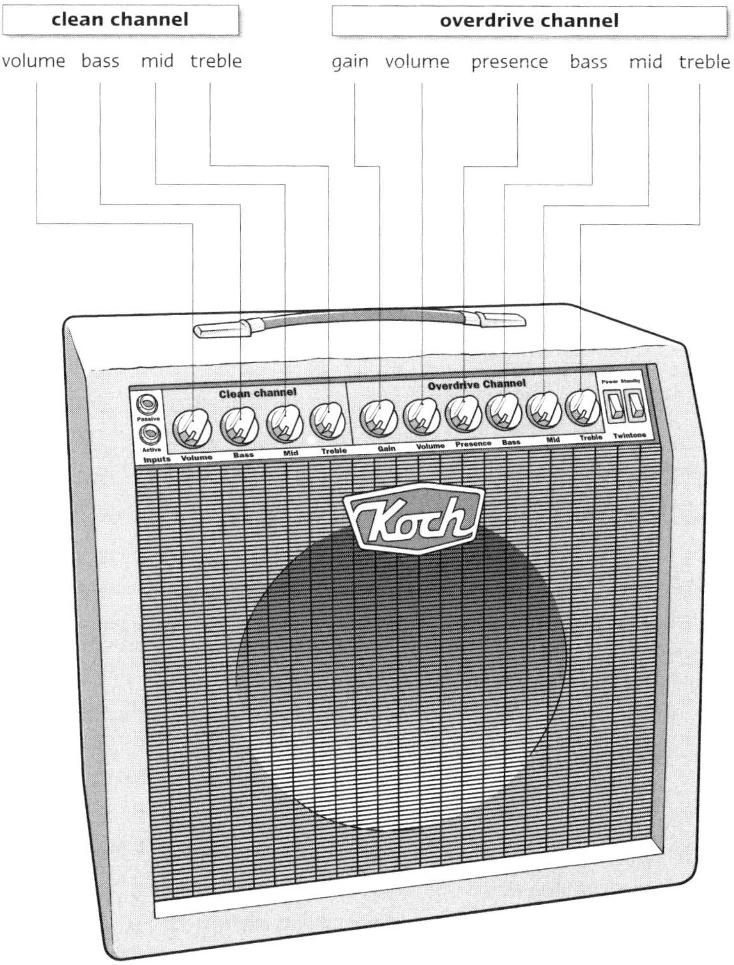
clean channel
volume bass mid treble
overdrive channel
gain volume presence bass mid treble
Clean channel
Overdrive Channel
Koch

Guitar amp with independent tone controls for each channel.

TIP

Harmonica amplifiers

Like electric guitarists, blues harp players like a distorted sound, and many of them use guitar amps. A few companies make dedicated harp or harmonica amplifiers, featuring tube amplification and a spring reverb (see page 119). More information on harmonica amplification can be found on various dedicated websites (see page 276).

BASS AMPS

Amplifying the low notes of a bass takes a lot of energy. That explains why the bass player almost always has a more powerful amp than the band's guitarist.

Highs too

Originally, bass amps were made to reproduce the deep, low, fundamental notes. But bass playing has evolved, and so have bass tone and bass amplification. Almost all of today's bass amps have a tweeter to deliver bright, percussive slap and pop sounds, and many have extensive midrange control to dial in your favorite tones.

ACOUSTIC AMPS

An acoustic instrument is an instrument that can be played without an amplifier. That definition makes the term 'acoustic amplifier' a bit weird — but it's here to stay.

More instruments

Acoustic amplifiers were originally designed for acoustic guitarists (that's where their name comes from), but they're used for other acoustic instruments as well, including violins, banjos, horns, and vocals. Most models feature a microphone input.

Feedback

When acoustic guitars and other string instruments are amplified, they're often bothered by *feedback*. Feedback is the loud, penetrating screech you hear when someone points a microphone at a speaker and, with acoustic guitars, it's often a low-pitched howling. Acoustic amps and double bass amps often have a *notch filter* that can filter out that horrific tone. A *phase switch* is another feature that helps fight feedback.

19

KEYBOARD AMPLIFIERS

Keyboard amps are designed for all keyboard instruments, but they're also used for horns, vocals, and other instruments.

Stereo channels

Keyboard players often use more than one instrument (e.g., a synth, a digital piano or organ, and sound modules) so keyboard amps typically have three or four channels. As these instruments are usually stereo, so are the channels. Each stereo channel has a level control that allows you to balance the various instruments or voices. The same channels can be used for the acoustic instruments mentioned above as well, provided that there is a microphone input.

Stereo

Notwithstanding their stereo channels, most keyboard combos have a mono amplifier. To get stereo, you need two of them. Such a setup can be used as a small sound system too, amplifying a keyboardist, two horn players and a singer, for example, while the guitarist and the bassist use their own amps.

SOUND SYSTEMS

Like a keyboard amp, a sound system or PA (*public address*) system is designed to reproduce any type of instrument, but it offers a lot more dedicated controls and options.

Small or large

Small sound systems are often used for acoustic instruments, keys, and singers only. Larger systems are also used to further amplify the output of guitar and bass amps, and for drums, just like the PA systems that are used in large venues and stadiums.

Basic setup

A basic sound system consists of a mixer, a power amp, and two speaker cabinets.

20

Mixer

The mixer has a number of channels, one for each instrument or microphone. Each channel has a level control, usually a sliding *fader* that allows you to literally 'mix' the volume levels of the channels to get a perfect balance between the instruments. Each channel also has its own gain (or *trim*), tone (EQ), and effect level controls. A separate *master section* controls the overall sound.

A basic mixer.

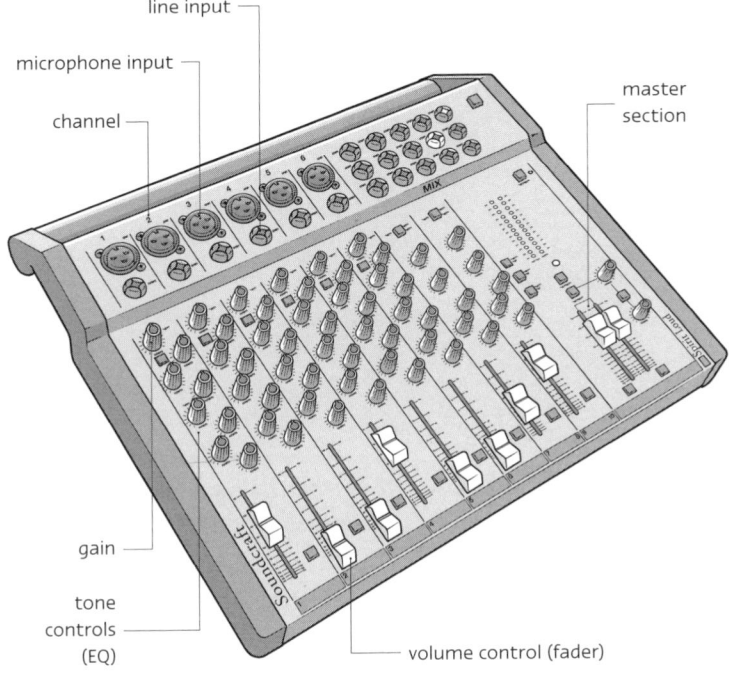

line input

microphone input

channel

master section

gain

tone controls (EQ)

volume control (fader)

Power amp

A mixer has a separate preamp for each channel. The power amp can be built into the mixer (*powered mixer*), the speakers can have built-in power amps (*powered speakers* or *active speakers*), or you can use a separate power amp.

Front-of-house

Sound systems usually use two- or three-way speaker cabinets. These are also called the *front-of-house* (FOH) speakers or *mains*.

21

Subwoofer

If the bass or drums are played over the sound system, their lowest sounds usually require a dedicated *subwoofer*, a cabinet with a special 15" or even an 18" low-end speaker.

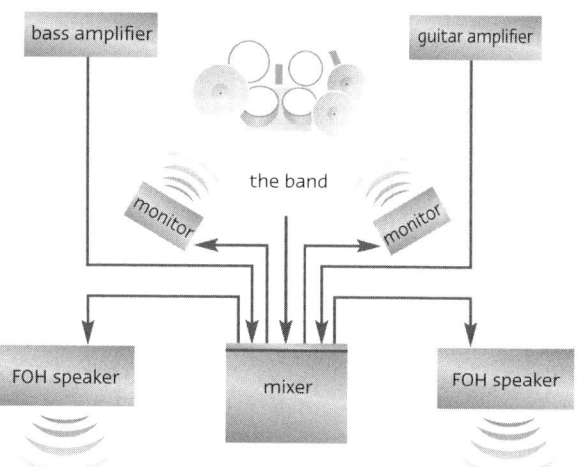

Monitors

Two speaker cabinets may suffice in small venues, but it's often best to add separate *monitor speakers* that are directed toward the band. They allow the musicians to hear themselves play or sing — yet not at the high volume level that is required for the mains.

Portable systems

Several companies offer portable, all-in-one sound systems that include a powered mixer, two speakers, speaker stands, cables, and sometimes even one or two microphones.

Multi-purpose

Mixers can also be used for home studio applications, to make a balanced recording of a rehearsal, and for many other purposes. The smallest ones have just two channels.

22

EFFECTS

Effects are used to add color and life to the sound. *Reverb*, for example, is widely used to provide ambience, warmth, and spaciousness. Many guitar amps, acoustic amps and keyboard amps have built-in reverb. *Chorus* is another popular effect. A chorus is used to make the sound a bit thicker and fuller, as if you're hearing a 'chorus' rather than a single vocalist. A *delay* repeats what you just played; a *pitch shifter* adds pitches to the notes you're playing; a *fuzz box* adds grit — and there are many more.

Tipcode AMPS-002

This Tipcode plays a brief audio sample on a digital piano six times: first dry (without effects), then using reverb, chorus, wah-wah, flanger, and delay respectively.

TIPCODE

Controls
Effects usually have a number of controls that allow you to dial in the sound you're after. If you're using an effect called *vibrato*, for example, the pitch of the notes you play moves up and down. The controls let you adjust how fast this happens, or how much the pitch goes up and down. These are two of the effect's *parameters*. Most effects have one or more adjustable parameters.

Built-in
Effects can be built into your amplifier or a mixer. This is a practical and fast solution (no messing with cables and plugs!), but you're not free to choose your own effects. Also, built-in effects usually can't be tweaked as precisely as those in dedicated 'outboard' effects units. That said, you can use the built-in effects as a start and add outboard units later on.

23

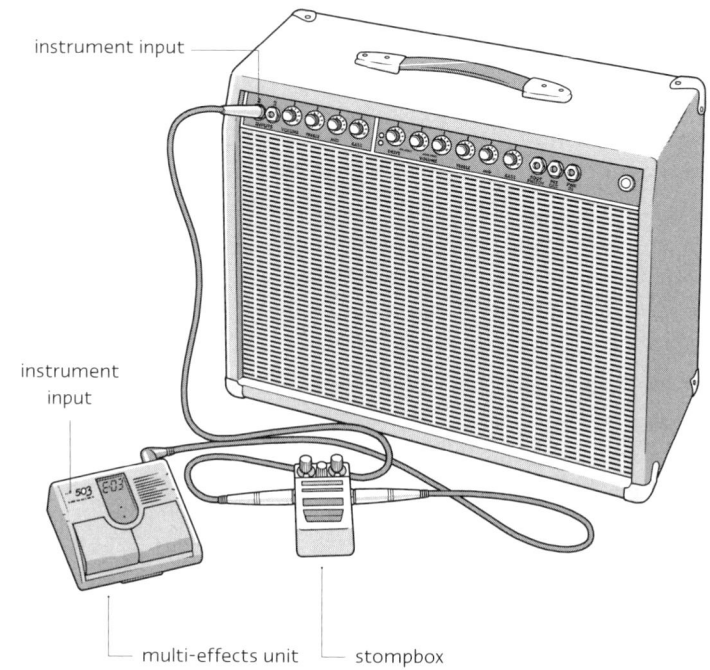

instrument input

instrument
input

multi-effects unit   stompbox

Stompboxes

Guitarists and bassists often use *effects pedals*, a.k.a. *stompboxes*:
small units that typically house one, and sometimes two or three
effects. You turn the effect(s) on and off by stomping on a rugged
footswitch, which may be hidden under a pad. A number of rotary
knobs are used to control the effect parameters.

Multi-effects

A *multi-effects unit* or *multiFX* processor is a larger unit that
features anywhere from two or three to over a hundred effects.
These units are programmable: Once you've set the parameters
for an effect, you can store the settings in a *patch* or *preset* that you
can recall at any time. There are multi-effects units for guitarists
and bassists, but also for vocalists and other musicians. Units
come in floor-mounted and desktop models.

Rack effects units

As mentioned before, effects can also be housed in 19" rack-mountable units. Rack effects are used in studios and PA systems, but many musicians have them in their stage rig too. Most of these units are operated with a foot controller that allows you to turn effects on and off, or to activate pre-programmed settings (*presets*).

Expression pedal

Floor-mounted effects units often have an *expression pedal* that you can rock back and forth to change one or more parameters in *real time*: apply more or less vibrato to the sound, or simply adjust your overall volume as you play.

Modeling preamps

There are a growing number of effects units that emulate the sound of famous bass and guitar amplifiers, cabinets, and stompboxes. This is known as *modeling*. These units double as a preamp: You can connect them directly to a power amp, a mixing board, or the sound card of your computer. In addition to these *modeling preamps*, there are *modeling amps* that have built-in amp and effect models in a combo amp or stack.

Tipcode AMPS-003
Here's a brief demonstration of four digital models of well-known classic guitar amplifiers, all produced by the same modeling amp.

 **TIPCODE**

Microphone modeling

A similar technique, *microphone modeling*, is used for singers and acoustic guitarists.

3

Buying Equipment

Amplifiers and effects come in various price ranges. How much should you spend, how much can you spend, and what's the best place to spend it? Here are some things to consider before you go out shopping.

There are packs that include both an amp and a bass or guitar, as well as a gig bag, a strap a pick, and an instructional dvd for less than a hundred fifty dollars — and you can easily spend twenty times that amount on a single amp.

Spending more

Generally speaking, spending more money buys you higher quality equipment: better parts, better construction, better sound, higher power ratings, and more features. You may not always want all of that, however. If you're looking for specific features in a keyboard amp, for example, you may find them only on models that have twice the amount of power you need, and a correspondingly higher price and weight.

Features or sound

... or you may have to choose between an amp that has lots of clever features and an equally or higher priced model that is extremely basic, but just sounds better. Use your ears to decide. Some great musicians go for very basic, utilitarian amps, while others prefer complex systems with numerous controls.

Size or sound

Likewise, there are impressive bass cabinets with no less than eight speakers and a tweeter that cost a mere six hundred dollars, and much smaller two-speaker cabinets that cost the same. Listen with your ears and not your eyes — it's all about the music.

PRICES

The prices listed in this book are mere indications of what you can expect to pay. While the prices of high-end products are quite stable, budget products seem to continuously offer more and more value for your money.

Three tips beforehand

The following three tips may sound quite obvious — but it's good to take note of them.

- Even **the most expensive amp** can't make up for a poor-quality instrument. That said, the resulting sound may just be what you've been looking for.

- Any instrument will only sound **as good as the amp** you buy.

- If you want to emulate the sound of **your favorite artist**, it may help to buy to the same equipment, or to focus on products from the same manufacturer. However, you still won't sound exactly the same. A huge part of anyone's sound is in the hands.

Guitar amps

The lowest priced guitar amps are meant for practicing only. If you're going to use your amp in a band, you need to spend more, say anywhere from three hundred dollars and up. If you're looking for a tube amp, be prepared to spend twice as much or more.

Boutique

The most expensive guitar amps often have a very limited number of features and controls. These are known as *boutique amps*, handmade in small numbers, using top-notch components throughout.

Bass amps

Practice amps start at less than a hundred dollars. Expect to pay around four to six hundred dollars for a bass combo amp that allows you to be heard in rehearsal and small venues. Powerful combo amps cost up to fifteen hundred dollars or more. You can also spend that kind of money on a single head, and reserve another fifteen or so for speakers.

Keyboard amps

Keyboard amp prices start at less than two hundred dollars for a 10 to 20 watt practice model. For three times that price you can get yourself a 100 watt combo that doubles as a sound system, big enough to play small venues — but you'll need two of them if you want stereo. High-end keyboard amps are well over a thousand dollars.

Acoustic amps

Acoustic amps range from around a hundred to more than a

29

thousand dollars. Five hundred dollars buys you a good-sounding amp with plenty of power for most situations.

Sound systems

Very basic sound systems, featuring two speakers and a powered mixer with four to six channels, are available for less than a thousand dollars. Don't expect it to fill a medium-sized hall with sound, but such a system can be sufficient to amplify a singer, one or two horn players, and a keyboardist, for example.

Effects

Stompboxes range from around thirty dollars to more than ten times that amount, depending on the type of effect and the quality of the product. By comparison, multi-effects units that offer a multitude of effects for a hundred to a few hundred dollars seem very affordable. That is not the only difference, of course. Chapter 13 tells you all about it.

TIP

Your favorite musician

If you'd love to reproduce the sound of your favorite musician, it will help to buy the same equipment, if possible — but do note that you can only sound the same if you can play just as well.

THE STORE

Because comparing sound quality is so important in selecting amps or effects units, you may prefer to go to a store that has a lot of equipment in stock. However, it may be easier to make your choice in a store with a small but representative selection that you can try out.

People and time

Small or large, one of the main things that makes for a good store

30

is that the people that work there know what they're talking about and enjoy their work.

Different stores

It's good to get to know amps and effects of different brands. So if possible, visit more than one or two stores. A number of brands will be represented in most music stores, which allows you to compare the different stories you will be told on identical pieces of equipment.

Testing gear

Getting plenty of time and opportunity to check equipment out is another important element. Some stores even have soundproof booths for that purpose, so you can play at the volume level you intend to use the amp at.

Take someone along

If you've just started playing, it'll be hard if not impossible to properly judge an amplifier or effect. Take an experienced player along when you go shopping, or go to a store where the staff is willing to demonstrate the products you're interested in. Ideally, they should show off the gear, not their own playing ability.

TIP

Essential

Of course, price can be important in deciding where to buy. But when it comes to musical instruments, customer support and service are often just as important.

Online and mail order

You can also buy your equipment online or by mail order. Unlike most musical instruments, there are usually no differences between identical amplifiers or effects units: The one you order online will sound the same as the one you heard onstage or the one your friend plays (with the possible exception of handmade boutique amps). Most online and mail-order companies offer a return service for most or all of their products — but you can't test and compare amps the way you can in a regular store.

31

Warranty

Final tip: Don't forget to ask around for a brand's reputation on customer support, and note that warranty regulations differ from one brand to the other. Some companies offer a one-year limited warranty on the amp, and three months on the speakers, for example; others simply exempt speakers and other vulnerable parts, such as tubes.

SECONDHAND

You can also buy used equipment. Musicians sell their amps because they get a better one or because they stop playing, they trade in their stompboxes for multi-effects processors (or the other way around), and so on. Expect to pay about half of the regular price for equipment that has been used for a couple of years, as long as it's in mint condition. Amps and effects can last for many years, unless they've been treated badly — and you can usually tell that from the way they look. Take that into account if you plan to sell equipment too, and also note that a manual and the original box can help raise to the price.

Privately or in a store?

Purchasing a used amp or effect from a private party may cost you less than buying the same unit from a store. One of the advantages of buying used equipment in a store, though, is that you can go back if you have questions. Also, music stores may offer you a limited warranty on your purchase. Another difference is that a good dealer won't usually ask an outrageous price, but private sellers might — because they don't know any better, or because they think that you don't.

Vintage equipment

Some musicians prefer the original classic, vintage amps and effects rather than modern equipment. If you're one of them, make sure that what you buy is not too old, and get an expert's opinion if you're not sure. Vintage amps and effects can be quite expensive.

Reissues

Because of the popularity of vintage gear, many companies make reissues. In some cases, these are exact copies of the old model. Others have been improved upon and include 'new' features. And many old-looking amps just have a vintage look, and nothing else.

MORE INFORMATION

If you want to know all there is to know, stock up on musicians' magazines that offer reviews of the latest gear, and on all the brochures and catalogs you can find. Besides containing a wealth of information, the latter are designed to make you want to spend more than you have — or have in mind — so ask for a price list. Of course, the Internet is great source for up-to-date product information, and there are many more books on the subject available. Titles, websites, and additional information can be found on pages 264–276.

4

Features

There are many features that are common to most amp types, and to some effect devices as well. These shared features are discussed in this chapter, which also includes basic information on speakers, digital equipment, and MIDI.

First, the manual. Some amps and multi-effects units (especially programmable ones) have so many features that not reading their manuals will probably rob you of half the things you paid for. But even basic amps and effects may have possibilities that you'll only find in the manual, so at least take a look at it.

Your safety, your technician
When reading a manual, don't skip the safety instructions. This book doesn't replace them! A schematic diagram of the electronic circuit may be included too. Save it for your technician in case you run into technical problems.

THE OUTSIDE

Most amplifiers are taken on the road, so they need to be rugged and preferably not too heavy.

The housing
Combo amps and speaker cabinets usually have a wooden enclosure. Birch plywood is a popular choice, but poplar and okume are much lighter. Wood thickness is usually around ¾"(19 mm).
Solid wood is rarely used; it is typically said to add some extra resonance and warmth to the sound. Many companies use MDF (fiberboard) or synthetic materials, from lightweight, environmentally friendly composite shells to reinforced polypropylene. The material of choice may have some influence on the sound, but it's more important that it is used to create a strong, stiff, acoustically inert frame for the amp and speakers.

Dimensions
Making a speaker cabinet is not simply a matter of building a box big enough to house the drivers. Many companies even use computer programs to calculate the dimensions of their enclosures, which have a direct result on the overall sound. Even the stiffness of the enclosure has its effect.

36

Covering

The outside is typically carpeted or covered with thick, strong vinyl that is commonly dubbed Tolex. Vinyl is easier to clean, but industrial-grade carpet is more scratch-resistant. Black is the standard 'color,' but some companies make amps and enclosures with vinyl, carpet, or genuine leather coverings in various colors, or with a see-through lacquered, hardwood exterior, among other variations.

Corners

The vulnerable corners are protected by metal or plastic corner caps. Plastic feels and looks smoother; metal looks tougher. Cabinets may have stackable corners or top indents that receive the feet or casters of stacked enclosures.

Feet

Feet are usually rubber, which can make it troublesome to push heavy gear into position. Metal feet, on the other hand, may leave scratches — but if you like the stage to boost your low end, they may help a little.

Grille

The speakers are protected either with strong, acoustically transparent grille cloth or with a metal grille. If the grille's grid is too wide, objects may still damage the speakers. A grille that's wedged in (to allow for close miking; see page 59) rather than screwed on, may come loose by itself. Grilles shouldn't resonate or rattle. A curved grille helps prevent denting. Metal grilles are also used to protect the vulnerable tubes or *valves* in tube amps.

Handles

The heavier the unit, the more important the handles will be. Heavy equipment preferably has a top handle as well as a pair of side handles. Some units have a bottom handle too. Well-placed handles allow for comfortable lifting and carrying, so check! Recessed handles, controls, and jack plates make things easier to handle, and they're less vulnerable.

37

Weight

Good amps used to be heavy due to their heavy speakers (with the magnets being the heavy part) and transformers, which can weigh up to ten pounds in powerful amps. Though new technology allows for high-quality yet lightweight components, a combo amp that's fit for rehearsals and gigs still easily weighs thirty pounds (14 kilos) or more, and a guitar speaker cabinet with four 12" speakers can be four times as heavy.

Casters

Heavy equipment is easier to move if it has wheels or *casters*, which are usually intended for smooth surfaces only. Stem-type removable casters can break or fall out, so some companies prefer plate-mounted casters. These are available in removable versions too. If the casters come with brakes, don't forget to use them.

Dolly-style

Some amps and cabinets have two casters and a removable or telescoping dolly-style handle to facilitate transport. There are models that also feature a quick-release lid to cover the speaker and controls. Other companies sell amplifier bags with built-in casters. More tips on transporting and protecting your equipment can be found in Chapter 16.

Tilting amps

Tilting a speaker cabinet or combo amp backwards directs the sound toward your ears or your audience's ears. Some cabs and amps have tilt-back legs, a tilt hinge, a spring-loaded handle, a lifter mechanism, or a kick stand for that purpose, or you can rock them back by detaching the rear casters. As an alternative, there are amp stands with various tilt positions. Tilting an amp makes you lose the low-end coupling to the floor, however, which is why some amps have an angled baffle board.

Wedge

Wedge-shaped enclosures allow you to position them at two or even three different angles, improving their projection. Angles vary widely per design, from 15° to 55° and up. Monitor speakers usually have a *wedge* format: They're floor-mounted, projecting their sound toward the band without blocking the audience's view. Various bass combos are designed to be used either straight up or 'kicked back.'

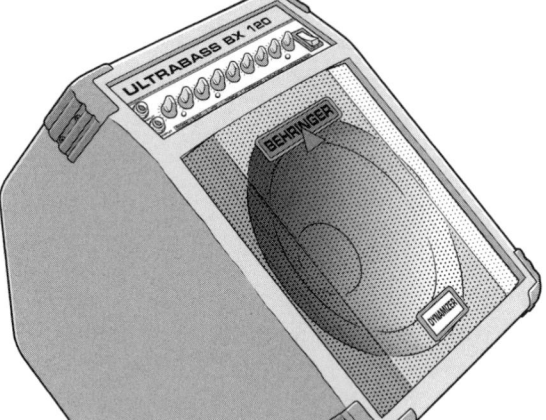

A kickback combo.

Speaker stands
PA speakers and some keyboard combos and acoustic amps are pole mountable so their sound reaches the people in the back as well. The adapters usually mount to standard 1 ⅜" (35 mm) speaker stands. Small (acoustic) combos and personal monitors can sometimes be mounted on microphone stands.

AC POWER

A detachable AC power cord makes an amp easier to handle and a broken cord can be instantly replaced, but it is a more expensive solution than fixing the cord inside the amp. The only advantage of a non-detachable power cord is that you can't forget to bring it. If the cord is non-detachable, see if the amp has *cable ears* to wrap it around for ease of transportation.

Three pins for a reason

Power cords with a three-prong plug should always be connected to a properly grounded AC outlet (see page 242–243).

TIP

> ## Switchable voltage
>
> For international gigs, the AC voltage needs to be switched from 110 (US) to 115 (Japan) or 230 (Europe) volts. Due to safety regulations, this is usually not user-switchable, if at all.

Extra outlet

A few amps have an AC outlet on the back (also called a *courtesy outlet*), so you can power another device.

CONTROLS

Most amplifiers have a dedicated knob for every job, and some have one or more rotary controls that double as push/pull switches. On digital equipment, each control can have a multitude of functions.

Top or front

The controls you're likely to use most often are usually mounted on the front or the top of the amp. Top-mounted controls make the amp more vulnerable to spilled drinks, and the same goes for top-mounted jacks. When the amp is on the floor, however, top-mounted controls don't force you to kneel down to see what you're doing. The most user-friendly amps have self-explanatory controls with easily readable texts and numbers, even on dimly lit stages (where small print can be illegible) or on bright stages (where the text on chrome panels can get washed out). Fantasy control names ('beef' for bass or 'shut up' for the headphone socket, for example) can make life more fun for some.

40

Quality control

The quality of the controls and switches, and the way they feel, can tell you a lot about the overall quality of the machine. Turning the knobs operates the *potentiometers* or *pots* on the other side of the panel, changing volume, tone, or anything else. Good pots make for knobs that rotate smoothly, but offer just enough resistance to prevent them from turning when you accidentally touch them. The knobs should not wiggle on their posts.

Chicken-heads and stoves

Rotary controls come in all kinds of shapes. Popular models include *chicken-head pointer knobs*, *stove knobs* and *skirted knobs*. Knobs may be one-piece affairs, or they have a separate cap or insert. *Tip:* If the cap or insert gets lost — and that does happen — you'll no longer be able to read the knob's position.

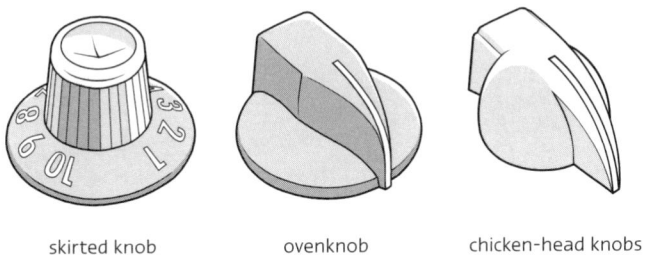

skirted knob   ovenknob   chicken-head knobs

Center-detented

Some knobs have a center-detented position, so you can feel where you are. You'll find this feature on balance controls and EQ controls, which are usually flat (*i.e.*, not affecting the sound) in their center position.

| Volume | Bass | Mid | Treble |

Volume at 11:00, bass at 12:00, midrange at 9:00, treble at 1:00.

41

Ranges

Volume controls usually range from 0 to 10 (or 11!), while EQ controls can range from 0 to 10, or from -12 to +12, for example. To indicate knob positions in a universal way, musicians often use o'clock positioning, e.g., 'At 2:00, the sound level is impressive.'

FOOT CONTROLLERS

Guitar amps usually come with a basic foot controller, *footswitch* or *floor pedal*. Most have two switches: one to switch channels, the other usually to turn the reverb or another built-in effect on or off. Similar pedals can be included with other types of amps as well. More complicated — and expensive — foot controllers are available as options.

A basic and an extended foot controller.

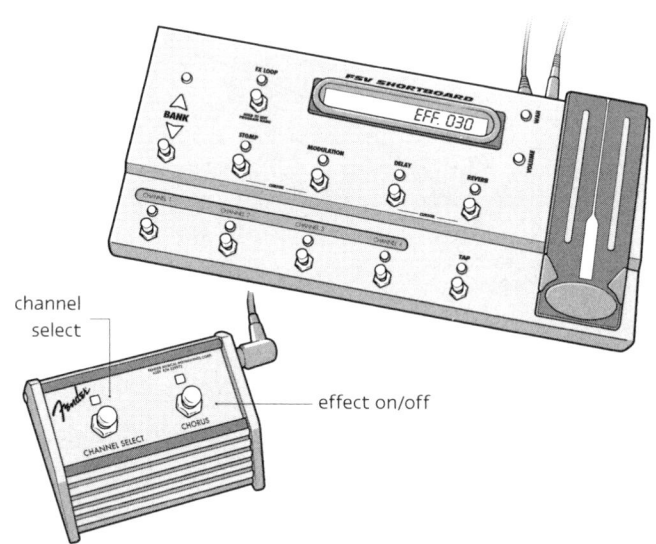

channel select

effect on/off

Functions

Extended foot controllers may allow you to set a tap tempo for delays or other effects (see page 177), to select presets, or to activate a built-in tuner or one or more effects, or any other feature that benefits from hands-free operation. These foot controllers can cost up to three hundred dollars or more. *Tip:* A *solo switch* switches between two preset volume levels. Hit the switch for that extra bit of power for your solo and change back afterwards.

Cable
As cables are vulnerable, it's best if the floor pedal has one that's easily replaceable. Some simple foot controllers use a basic instrument cable, so you're likely to have spare ones around. Others use a cable with special connectors. If so, it's wise to invest in a spare.

Spacing
If you have large feet, a relatively wide switch spacing will help prevent you from hitting more switches than you plan to.

(Non-)latching
Most footswitches are latching switches: Like a light switch, you use it to turn things on or off. Non-latching switches are like door buzzers: They're on as long as you hold them down.

EQUALIZERS

An equalizer is a powerful tone-shaping device. It can be used to even out things you don't like in the instrument's tone or the room's acoustics, for example, and guitarists typically use it to create their own sound. Learning how to tweak these tone controls takes time, practice, and patience, even if it's basic three-band EQ.

Suggested settings
The large variety of EQs (and amps, and instruments) doesn't allow for suggested settings in a general book like this, but manuals often list a number of sample settings, offering you some starting points. If you find an EQ setting you like, write it down. Also, check the tips under *EQ panel diagrams* on page 279.

43

Sound

Every note you play is made up of a number of frequencies, as you can read on pages 88–90. An equalizer allows you to influence the relative volume of certain frequency ranges, and that's what influences the timbre of the sound.

Room acoustics or timbre

An important function of the EQ is to adapt the sound to the room acoustics: Boost the treble range if the room sounds dead or wooly, or reduce the bass if the room is boomy. On keyboard and acoustic amps, EQs are mostly used for this purpose. On guitar amps, the EQ is primarily used to color and shape the sound.

TIPCODE

Tipcode AMPS-004
This Tipcode is a brief demonstration of various EQ settings for jazz, blues, and metal.

Rotary and graphic

Equalizers with rotary controls are known as *rotary* equalizers. They usually have two to five bands. Equalizers with more bands often use faders that graphically show you the setting — hence their name, *graphic* equalizers. Graphic EQs on amps usually have five to ten bands.

TIPCODE

Tipcode AMPS-005
For this Tipcode, we recorded a brief piano sample with different EQ settings.

44

TIP

More controls

More expensive amps tend to have a larger range of EQ controls. However, the number of controls has no bearing on an amp's quality: There are expensive guitar amps that have no more than a bass and a treble control. Most amps have a three-band rotary EQ, with controls for bass, mid, and treble.

Bass

Boosting the bass makes music sound warm, loud and fat. Turning the bass level down helps reduce rumble and promote clarity.

Mid

The midrange control can be used to boost important frequencies of vocals, horns, and many other instruments, emphasizing them and enhancing their clarity without drastically altering their sound. Extreme settings can make for a 'hollow' sound (by cutting mids) or a harsh and fatiguing tone (by radically boosting mids). As this frequency range is so important, many amps have two midrange controls: one for low mids, and one for high mids.

Treble

Opening up the treble control adds crispness to the sound. Opening it up too much can create hiss. Turning this control down cuts high frequencies, which may help to reduce the effect of high-frequency consonants, such as the sibilant 's' sound.

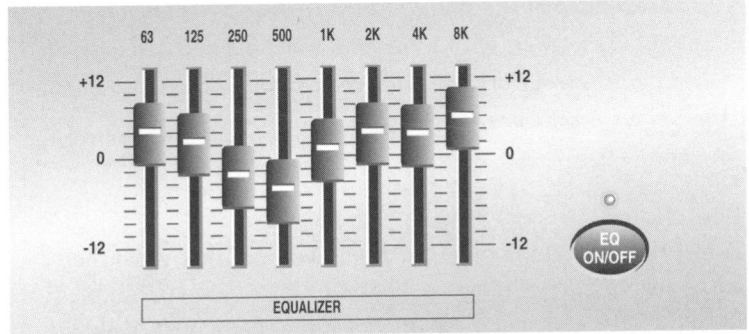

An 8-band graphic equalizer in a scooped-mid setting (see page 46).

45

Which frequencies

On graphic EQs, the frequency ranges of the various faders are printed on the panel. For example, a ten-band EQ typically has faders marked 31 – 63 – 125 – 250 – 500 – 1k – 2k– 4k – 8k – 16k respectively. The first fader then affects frequencies in the 31 hertz range; the highest is active in the 16 kilohertz band (see pages 88–90 if this doesn't make sense to you).

Octaves

If you look at the numbers in the previous paragraph, you'll see that the frequency range is doubled for each consecutive fader. That means that each consecutive fader controls a range that's an octave higher than the previous one. Studios and professional sound systems use 30- or 31-band equalizers, with each band covering no more than ⅓ of an octave. This allows for maximum control over the frequencies you want to cut or boost.

Volume control?!

Tip: When the EQ's controls are in their middle or flat position, they don't affect the sound. If you move all of them up or down the same amount, you will hardly hear the timbre change. Instead, the sound will get louder or softer: Boosting or cutting all frequency ranges the same amount has virtually the same effect as turning the volume control up or down.

TIP

Scooped mids

One of the advantages of a graphic EQ, as said before, is that you can clearly see what you're doing: It provides a detailed, graphic image of the sound. Terms such as a 'scooped mid sound' for settings with a cut midrange and boosted lows and highs, stem from the relative position of the EQ's faders, as shown on the previous page.

The choice

Hopefully, your amp's manufacturer has carefully chosen the adjustable frequency ranges so that they have the maximum effect. Both the instrument and the type of sound the amp is made for play an important role in that process. On some guitar amps, the bands even vary depending on which channel you use: The clean channel's EQ controls affect different bands than the lead channel's.

Parametric equalizer

Some equalizers allow you to choose and adjust the midrange frequency band yourself. These are equalizers with a *parametric midrange*, featuring three controls for this one range:

1. The first control allows you to set much you want to boost or cut the midrange.

2. With the second control you can set that midrange from, say, 800Hz to 3kHz.

3. A third control, labeled *Q* or *slope*, enables you to also adjust the width of the range (the *bandwidth*) you want to affect.

TIP

Fixed

On some amps, you have a choice of two fixed Qs, usually labeled 'narrow' and 'wide', and/or a choice of two fixed frequency bands, rather than having a whole range to choose from. This makes a parametric EQ easier to use, of course.

Semi, quasi, sweepable

A *semi-parametric EQ* has separate level and range controls, but

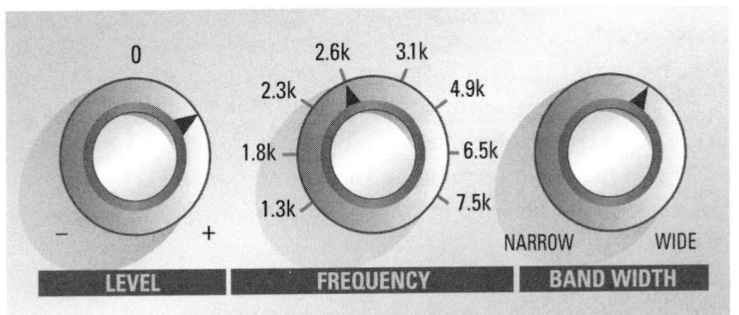

| LEVEL | FREQUENCY | BAND WIDTH |

A parametric EQ allows you to set the EQ level, the frequency range, and the bandwidth.

47

no Q control. Semi-parametric EQs are also known as *quasi-parametric*, *sweepable*, *sweep* or *swept EQs*. If there are separate sets of controls for the low and the high midrange, they may be labeled *low paramid* and *high paramid*, for example.

Time

Learning to use a parametric or semi-parametric EQ, again, takes time and practice. A good way to get started is to set the level control fully clockwise or counterclockwise, and then slowly sweep the frequency control. This way you can easily find the frequency range that needs to be cut or boosted — and adjust the level accordingly. *Tip:* With the level control in its center detent position, the range control has no effect at all!

Shelves, bells, and peaks

Three more terms that you may come across: A *shelving EQ* means that all frequencies below or above a point will be affected. On many equalizers, the bass and treble are low and high shelving EQs respectively. The mid control is a *peaking EQ* or *bell-shape EQ*. Opening it creates a bell-shaped 'peak' at the selected frequency. Both shelving and peaking EQs are shown below.

A shelving low (cut) and high (boosted) EQ, and a (boosted) peak mid EQ.

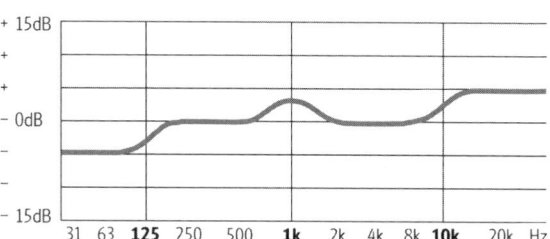

How much

How much an EQ control affects a certain range is expressed in *decibels* (*dB*; see page 81). This is usually indicated on both extremes of the rotary controls. If it says -15 on one end and +15dB on the other, turning the knob fully counterclockwise cuts the range 15 decibels and opening it up provides a 15dB boost.

Active and passive

EQs that can both cut or boost frequencies are known as *active*

48

EQs. The controls are flat in their center position, at 12:00. Passive EQ controls, which only cut frequencies, usually go from 0 to 10, like a volume control.

Higher is more

When testing amps, trust your ears to find out how sensitive or powerful the EQ is, and how responsive the amp is to your settings. A tip: When using the EQ, adjust the treble first and the bass last. If you do it the other way around, a boosted bass often requires more mids, and more mids might require a higher treble setting — so you'll end up playing a lot louder than you planned to.

EQ panel diagrams

Another tip: If you've dialed in a perfect sound on your own amp, write down the settings. Some manuals provide you with blank *EQ panel diagrams* for that purpose, next to a number of sample diagrams. You can also make those yourself.

Alternatively, you can mark your settings on a white strip of tape that you can stick under the controls, provided there's room to do so. If your amp has user presets, you can simply store your settings (see page 206).

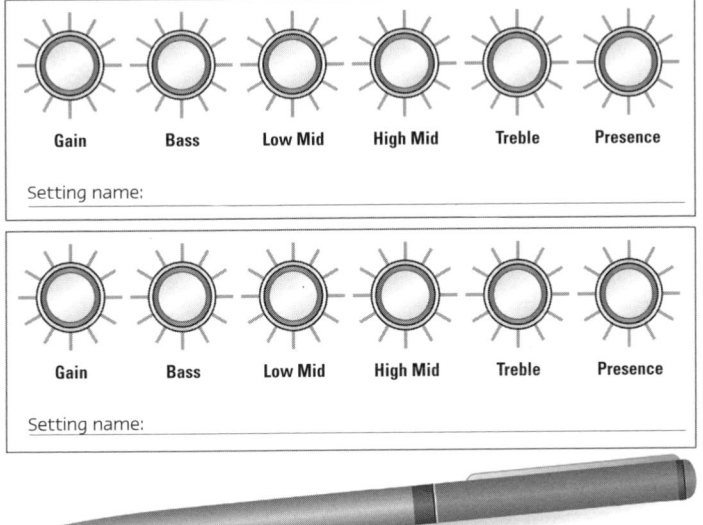

Record your settings with blank EQ panel diagrams.

More EQ controls

Many amps have additional tone controls, such as switches that boost low, mid, or high ranges, or a rotary control that produces a

Tipcode AMPS-006
This Tipcode demonstrates the effect of various shape settings on the sound of an electric bass guitar.

Tips

- To learn how to really use an EQ, you need to know the role that various **sound systems** play in an instrument's timbre. For example, a bass drum is a low-sounding instrument with a lot of 'oomph' around 80 to 100Hz, but to enhance its definition in the mix, increasing the lows won't help: Instead, you will need a boost in the 2.5 to 6kHz range. Likewise, boosting the treble will only increase hiss and noise if those high frequencies aren't present in the instrument's tone. Dealing with these subjects in more detail goes beyond the scope of this book, however.

- Boosting one or more frequency ranges too much may result in a distorted sound or speaker damage. A few amps have an EQ clip light that warns you if this is about to happen.

- Use your amp's EQ in conjunction with the EQ controls on your guitar, bass, or keyboard: The tonal possibilities are endless.

- And if that's not enough, you may want to get yourself a stand-alone equalizer (see page 184).

variety of preset tonal adjustments that add to what the regular EQ controls can do. Such controls are usually labeled *shape, contour, enhancer,* or *voicing.* These controls often boost and cut ranges other than the ones you control with the built-in equalizer.

INS, OUTS, AND PLUGS

The number of inputs and outputs tells you something about an amp's versatility. A really versatile combo amp has a headphone jack for silent practice, an extra speaker output, a special output to connect the amp to a sound system or a recording console, separate connectors for effects and footswitches, and sometimes even more connectors.

Front, rear, or top
Most amps have front-mounted jacks. Others have all or some of the lesser-used jacks mounted on the rear. Rear-mounted jacks may look a bit neater, but front-mounted jacks are commonly easier to use. Top-mounted jacks, again, are most vulnerable for spilled drinks. Make sure the connected cables don't get in the way of the controls. *Tip:* On rack equipment it may be easier to have most connectors on the rear panel.

Quality
It's hard, if not impossible, to assess the quality of the jacks, but it's good to know that there are differences. Good jacks have a better, longer-lasting hold on your plugs and they are less likely to cause hum or drop-outs. The combined weight of a plug and a cable shouldn't put too much stress on a jack. A cable that is jerked around a lot certainly does, so be careful.

Phone jacks
Most instruments use *instrument cables* with ¼" *phone plugs* that fit the matching, small round phone jacks on the amp. They're usually mono. Stereo instruments such as keyboards simply use two mono cables and two mono inputs — one for the right channel, one for the left channel.

51

Stereo phone jacks

Headphones and other stereo devices have a stereo plug, which requires a stereo jack. Stereo plugs are easily recognizable, with two narrow (usually black) bands that separate their three contact points: tip, ring, and sleeve (TRS). They're also known as *TRS* plugs. Mono plugs (TS) have a single band.

Mini

Headphones usually have a 3.5mm mini-plug with an adapter that allows it to be connected to a ¼"(6.35 mm) phone jack. Mini-plugs are available in both mono and stereo.

Various types of plugs.

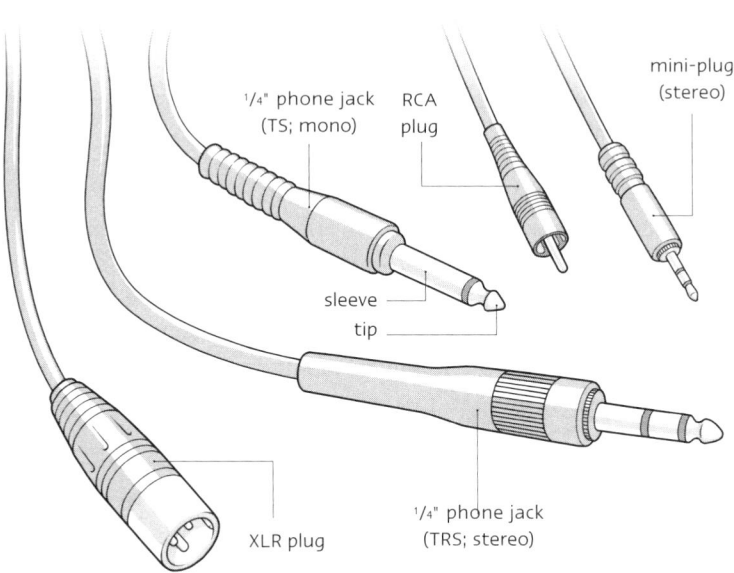

¼" phone jack (TS; mono)

RCA plug

mini-plug (stereo)

sleeve

tip

XLR plug

¼" phone jack (TRS; stereo)

Plug = switch

In some cases, a plug acts as an additional switch as soon as it is plugged in. For example, plugging in a headphone jack may switch off the speakers. If so, this may not function properly if you use a mono plug instead of a stereo one, or vice versa. Check your manual.

Hole plugs

At high volume levels, you may hear air rushing in and out of unused jacks, caused by the movement of the speaker's cone. This

52

can be stopped with a set of *hole plugs*. Some amp makers prevent this type of noise by locating the jacks over a sealed chamber.

XLR connectors

Professional microphones are usually connected with *XLR connectors* or *cannon connectors*. The male version, used where a signal leaves a device or the cable, has three pins. Female XLR connectors, with three matching holes, are used where the signal enters a device or the cable. This is different from phone jack connections, where both inputs and outputs are female.

out of the mic (*male*)   out of the cable   into the mixer
                                               (*male*)   (*female*)

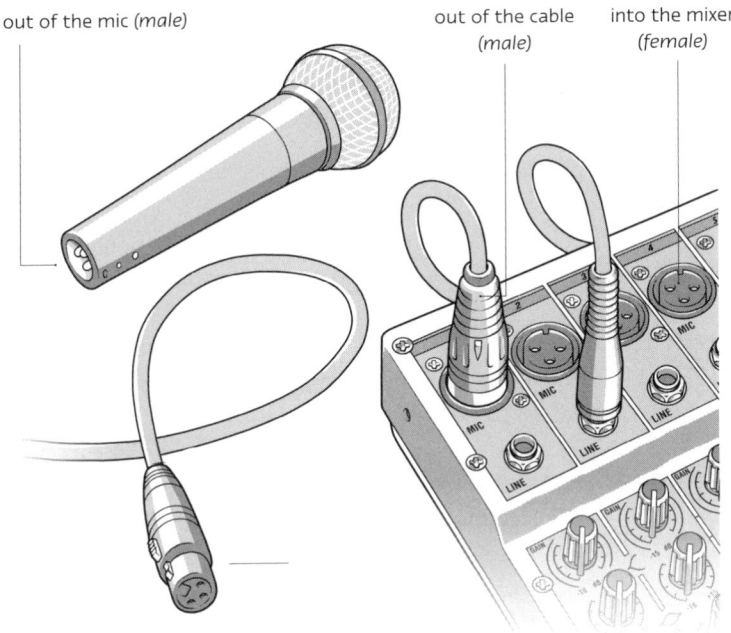

XLR uses male and female connectors.

Balanced connections

XLR connectors are typically used for so-called *balanced* or *symmetrical connections*. This type of connection uses two symmetrical conductors. In non-technical terms, the second conductor cancels out interference that was picked up by the first. As microphone signals are very sensitive to interference, microphones often use a balanced connection.

53

Unbalanced to balanced

Long instrument cables (18 feet and up) are more sensitive to interference too. To prevent interference, you can turn the unbalanced instrument signal into a balanced signal by using a DI box (see page 60). This requires the use of a balanced cable out of the DI box.

TIP

> ### Unbalanced XLR
>
> XLR is also used for unbalanced connections. For example, speaker cables with XLR plugs are unbalanced, using only two of the three pins. Conversely, balanced connections can also be made using TRS plugs, but this is less common.

Adapter?

Some devices have a balanced ¼" microphone input only. If your microphone cable has XLR plugs, use an XLR-to-¼" cable rather than using an *XLR-to-¼" adapter*: Adapters make the signal path more hazardous, and their weight puts extra strain on the jack.

Other uses

XLR and ¼" connectors are used for other purposes too, from connecting speakers to hooking up amps or effects.

Locking sockets and combined sockets

Tip: Locking XLR and phone plug sockets lock in the plug upon

Combined sockets accept both XLR and phone plugs.

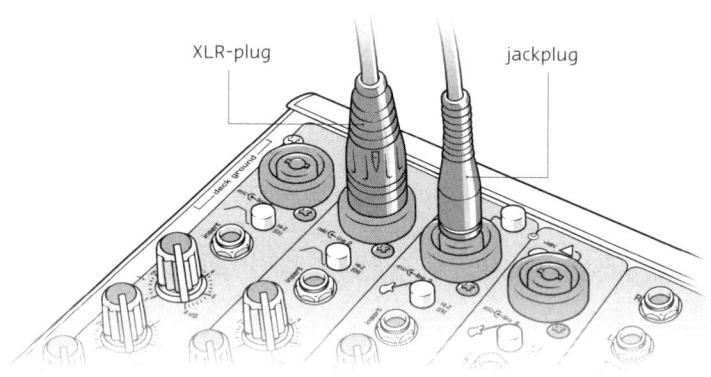

insertion. A push tab releases it. They're rare on combo amps. Another tip: There are space-saving *combined sockets* that accept both XLR and phone plugs.

CD players and other sources

Some amps have a special *auxiliary, aux* or *'jam along'* input for a CD player or a comparable sound source, allowing you to play along with pre-recorded music. The same input can be used for a keyboard or a drum machine as well. It may have ¼" phone jacks, mini jacks, or the same RCA sockets that your home stereo has.

Confusing

RCA plugs are also known as *phono plugs*. Confusingly, many people use this term for ¼" *phone* plugs too.

Speaker out

Amps often have an output for an extension speaker cabinet. This makes for a broader sound and may provide additional power, as you can read on page 84.

Speaker connections

Speakers should be connected with dedicated speaker cable, as described in Chapter 15. Amps and speaker cabinets usually have ¼" jack, XLR, and/or dedicated *Speakon* connectors. The Speakons have a number of advantages:

- You **cannot touch** the conductors (prevents both hum and electric shock from high-powered amps).

- They're **rugged**.

- They offer a **stable connection** surface.

- They can't be **unplugged accidentally**.

TIP

> ### Bananas
> *Some amps still use banana plugs, or they have speaker terminals that are designed to receive the cable's bare-wire ends, like the terminals on most home stereo amps.*

55

- They're **airtight**, preventing pumping sounds (see page 130).

- They can handle **heavy cables**.

- They have excellent **strain relief**.

Speakon plug and socket.

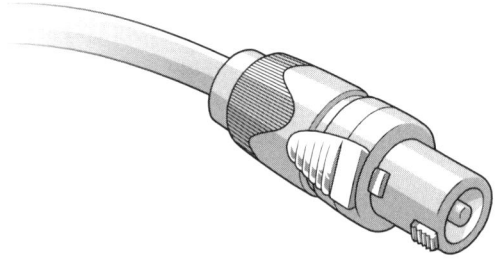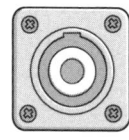

Headphone jack

Plugging in your headphones doesn't always automatically switch off the speaker(s). Why? Because the headphone output can often be used to connect the amp to a mixing board or a recording system, and in those cases you probably still want to hear the built-in speaker. Amps that don't automatically disconnect the speakers either have a switch to do so, or the master volume can be turned down without affecting the headphone level.

Speaker simulation

Distorted guitar sounds often sound very thin and scratchy when using headphones. If they don't, the headphone jack probably has

Tips

- *The more expensive the amp, the less likely it is to have a headphone jack, as these amps are less likely to be used for practicing.*

- *Never turn on a tube amp with a disconnected speaker (see page 117).*

- *If you're playing late at night using your headphones, make sure the speaker is really off!*

a special filter that mimics the sound of a guitar speaker (usually a cabinet with two or four 12" speakers). This is known as *speaker simulation*, *speaker compensation* or *cab emulation*. Such filters can also be found on other outputs and devices (e.g., DI-boxes and preamps), for the same purpose.

Which headphones?

You can use your regular headphones. Here are three tips:

- Models with an impedance lower than 75 ohms may **sound so loud** that you'll damage your hearing.

- For distorted guitar sounds, some players prefer to use really **cheap headphones**; good headphones may sound too clean.

- Bassists commonly prefer headphones with relatively **large drivers**, which improve the reproduction of their lowest notes.

Not too loud

Avoid high volume settings to prevent hearing and headphone damage. Always turn the volume level down before you put on your headphones, and gradually increase the volume as you start playing. Turn the volume down as soon as you hear a clicking sound, if playing with headphones makes you feel tired or if it causes ringing ears (and read pages 83–84!) *Tip:* Many amps have a separate level control for the headphone jack.

To tuner

Many bass amps and acoustic amps have an output labeled *tuner* or *to tuner*. It allows you to keep your tuner connected at all times without interfering with the signal path — and the shorter and more direct your connections are, the better.

Two tips

- A **tuner output** usually doubles as a line out.

- On bass amps, the tuner output may have a filter that cancels the bass' harmonic frequencies (see pages 88–90) that make it hard for the tuner to 'hear' the fundamental tone. This makes tuning **faster and more accurate**.

57

Mute switch

Plugging your instrument in or out may cause a loud pop, unless you turn all volume controls down. An *input mute switch* helps prevent these percussive noises. The switch should also mute the amp's line out. If it doesn't, and the amp is connected to a sound system, you will still hear the pop coming out of the system's speakers. *Tip:* There are special cables, plugs (e.g., Snap Jacks) and retrofittable items that help prevent those unwanted effects, allowing for silent instrument switches.

GAIN CONTROL

Most guitarists use their gain control to get varying degrees of distortion. Chapter 7 tells you more. Other musicians use it to match the amp's input to their instrument's output.
If this input control is set too low, the amp doesn't get enough signal. Turning up the master volume control helps, but will also boost unwanted hum or hiss. If the gain is set too high, the preamp gets too much signal. This will make it *clip*, resulting in unwanted distortion.

Clip indicator

A *preamp clip indicator* or *clip light* is a great help to set the gain

The indicator lights up when gain is set too high.

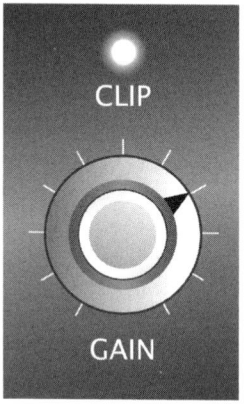

control. Turn the master volume down. Play as loud as you normally play while slowly turning the gain control up until the clip light begins to flash. Now back off the gain control to the point where the indicator lights on loud peaks only, then open up the master volume control to the desired volume level.

AMPLIFYING THE AMPS

In larger venues, the keyboards, vocals, and other instruments will run direct into the PA system — but guitarists and most bassists can't do without the sound of their own amp. The solution is simple: You don't connect their instruments to the PA, but you use the PA to amplify their amps. An amp can either be miked, or connected directly to the mixing board.

Miking
When miking an amp, the choice of microphones and their exact placement is crucial to the resulting sound: not only the distance between the microphone and the speakers comes into play, but microphones can be placed off-axis or on-axis, they can be aimed at the center or at the edge of the speaker, and so on. *Tip:* Some amps have removable grilles, allowing for very close miking. The closer the mic is to the speaker, the denser and less ambient the sound will be.

Line out or DI
While some can't live without the many options that miking an amp offers, others find it much easier to connect their amp directly to the PA. For this purpose, many amps have a *line out* or a *direct injection (DI) output*, also known as a *direct out*.

Inject
This output 'directly injects' the sound into a PA system or any other outboard equipment, from recording consoles to additional power amps. On some amps, the line or DI output level can be adjusted.

59

The difference

A DI is always balanced and typically uses XLR connectors; line outs commonly have unbalanced jack connectors. Some companies use different names for these outputs, such as *recording out* or *auxiliary out*.

Speaker emulation

To make distorted guitar tones sound really good over the sound system or on the recording, the DI or line out needs to be equipped with a speaker simulation filter — just like the headphone jack on pages 56–57. If there's no such filter on your amp's DI, you can use a separate *DI box* or *direct box* that has one.

DI boxes

The main purpose of a DI box, however, is to turn an unbalanced instrument signal (such as a guitar signal) into a balanced signal, which can travel long distances — think large stages — without picking up unwanted noise, such as radio signals or ground hums.

Active or passive

There are *active DI boxes* with additional features like a built-in preamp, tone controls, and a notch filter to help fight feedback. If you just want to send a balanced signal to the sound system or recording console, you can get by with a good *passive DI box*, which doesn't use batteries or a potentially hum-inducing adapter.

TIP

Tips

- They look deceivingly simple, so why spend some extra money in DI boxes? Because they do come in different qualities, just like all other equipment.

- Active DI boxes can be powered by phantom power (see page 164).

Doubling

As mentioned before, the headphone output can sometimes double as an unbalanced line out, unless its signal is too strong. Other outputs may be usable in more than one way too, so always check your manual. Combined outputs help lower production costs; dedicated outputs make for more flexible — and more expensive — equipment.

Ground lift

When connecting amps to PA systems and other gear, you may hear a *ground hum*. This can be solved with a *ground lift switch*, *ground lifter* or *earth lift switch*, commonly found on devices with a DI output. If there's no ground lift switch available, please check pages 242–243 for additional information. **Never** remove the ground pin of a three-prong power plug, or connect your equipment to a non-grounded power outlet; you risk electric shock if you do, as your manual will tell you.

balanced output (to PA)

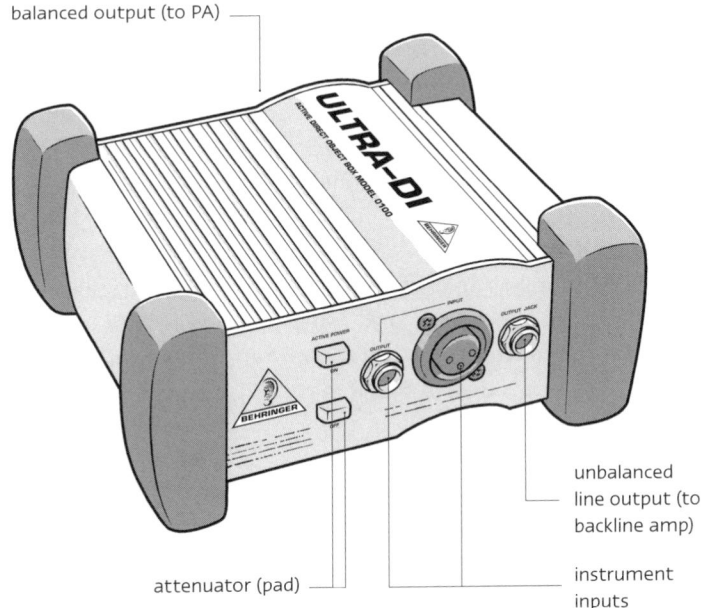

A basic active DI box.

unbalanced line output (to backline amp)

instrument inputs

attenuator (pad)

Post or pre?

A DI or line out can be *pre* (before) or *post* (after) EQ. If it is pre EQ, your tone control settings will have no influence on the sound that goes into your recording device or a second amp, for example. If it is post EQ, sound engineers have to deal with your EQ

61

settings, which they may not like. The perfect solution? An amp with a *pre/post EQ switch*.

Output select

Amps with an *output select* allow you to route certain signals. For example, you may have a click track (metronome) on one of your inputs that you want to route to your headphones or your monitor, but not to the PA or recording system.

EFFECTS LOOPS

If guitarists connect their stompbox effects between their guitar and their amp, the effects will always be affected by the preamp's distortion and its EQ settings. After all, the effect is part of the input signal, and the entire signal travels through the amp. Some effects (see page 195–196) are more effective or they just sound different when they are inserted *after* the preamp and the EQ. This is what an *effects loop* is for. Many amps have one.

Send and return

An effects loop has an output that sends the signal from the preamp to the effect. This jack, usually labeled *send* or *to FX*, connects to the effect's input. From the effect's output a second cable returns the effected signal to the amp, using a jack labeled *return* or *from FX*.

DI with level control, ground (GND) lift switch, and pre/post switch (see page 62).

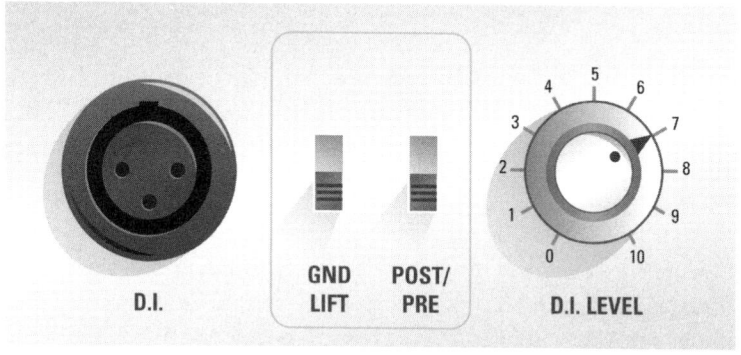

Serial and parallel

There are two types of effects loops.

- In a **serial loop** the entire signal leaves the preamp, it passes through the effect(s) and then returns to the power amp.

- A **parallel loop** has two signal paths: one that passes through the effect(s), and a parallel one that goes straight from the preamp to the power amp, bypassing the loop. At the return, the affected (*wet*) and the non-affected (*dry*) signal come together.

Parallel loops

Amps with a parallel loop have a *mix* or *blend control* to balance the *wet* signal (that ran through the effect) with the *dry* signal (that did not). In its middle position, you'll hear a balanced mix of both signals. With the blend control turned fully counterclockwise, the effect isn't noticeable at all. Turned fully clockwise, all you hear is the wet signal from the effect device — as if it were a serial loop.

Tips

- On some amps, the blend control doesn't allow for a 100% wet sound. Instead, the maximum setting is, say, 50% wet and 50% dry.

- Parallel loops are sometimes referred to as side chain loops or mixing loops.

Why?

Many effects units compromise the original signal to some extent, reducing low end, treble, depth of sound, and dynamics, while adding noise. A parallel effects loop reduces this effect as it allows you to add a bit of the original dry signal to the 'deteriorated' wet signal.

Serial loops

A serial loop is used for effects that need to process the entire

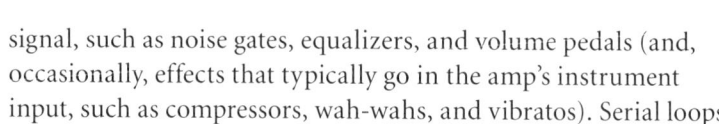

signal, such as noise gates, equalizers, and volume pedals (and, occasionally, effects that typically go in the amp's instrument input, such as compressors, wah-wahs, and vibratos). Serial loops are also referred to as *inserts*.

Parallel or serial?

If you're not sure whether you're dealing with a parallel or a serial loop, first check for a blend or mix control on the amp. This

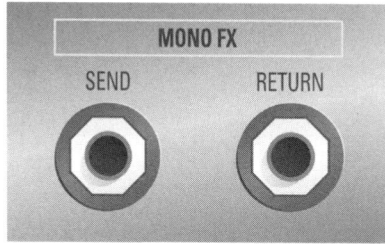

Send and return jacks.

control indicates a parallel loop. If the sound stops when you insert a plug in the return jack, it's a serial loop.

Parallel and serial

Tip: *Some amps have both serial and parallel effects loops, and others have a switchable loop, or even two.*

Preamp out, power amp in

Rather then using send and return labels, some companies label their serial loop jacks *preamp out* and *power amp in*. This shows that you can also use the preamp out to feed into another power amp, and that you can connect another (modeling) preamp to your power amp.

Main amp out, external amp in

A few amps can be daisy-chained for extra power: They will have sockets labeled *main amp out* and *external amp in*, for example. These sockets can be easily confused with the aforementioned preamp out and power amp in sockets. Don't.

64

More features

Here are some extra features that you may come across.

- Some effects loops have their **own send and/or return level controls**.

- Most effects units also have **level controls**, allowing you to adjust the level so that your volume doesn't go up or down when switching the effect in or out. On other units, you use this control to set the balance between the wet and the dry signal.

- Stereo effects will only be stereo if you have a **stereo amp** with a stereo loop. Some amps have both stereo and mono loops.

- A **footswitch** to switch all the effects in the loop in or out simultaneously.

- Some guitar amps have **separate loops** for each channel, so you can use different effects for your clean and distorted sounds.

- You can use the send jack as an **unbalanced line out**. The return may double as an input for line-level signals from a drum machine or a CD player, for example. If you want to jam along with a CD, the blend control of a parallel loop allows you to balance the volume levels of your instrument and the CD.

- Not every loop can deal with instrument-level signals from stompboxes and similar sources. **Some loops are switchable** between instrument-level signals and the line-level signals provided by professional 19" units (see page 86). Others accept both.

- A **buffered effects loop** prevents impedance mismatches.

TUBE, SOLID STATE, HYBRID AND MODELING

Bass amps, acoustic amps, keyboard amps and sound systems predominantly use the same solid-state technology (ss) that can be found in TV sets and other home electronics. Guitarists (and some bassists) often prefer amps that use old-style vacuum tubes instead. Why? Tube amps are heavier, more expensive and more vulnerable, and tubes age so they need to be replaced from time to time — but they do produce a warmer, smoother, more natural type of sound, especially when overdriven.

Vacuum tubes in a tube amp (Hughes & Kettner)

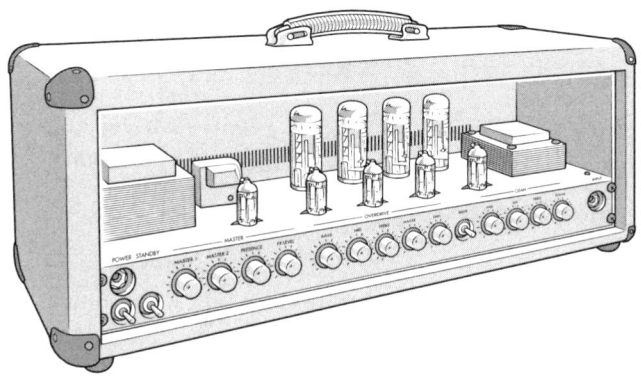

Preamps, effects, and hybrid amps
Tubes are also used in separate preamps and in various types of effects, adding sweetness and warmth to the sound. *Hybrid amps* combine a tube preamp with a solid-state power amp, or the other way around.

Tubes?
As tubes have such a reputation for good tone, lots of amps and effect devices have the word 'tube' or 'valve' in the name, even though there's no tube in sight.

MOSFET transistors
Some companies use MOSFET transistors, which some consider the solid-state equivalent of tubes.

Modeling and digital amps
Modeling amps use software to emulate the sound and character of a number of classic amplifiers. They're mostly used by guitarists. As modeling uses digital technology, modeling amps are often referred to as *digital amps*.

66

TIP

> **Class D**
>
> Class D amplifiers are sometimes referred to as digital amps, but the D does not stand for digital (see page 126).

DIGITAL EQUIPMENT

A growing number of amps and effects use digital technology. They may be entirely or largely digital, and there are tube amps with digital components too, combining old and new technologies.

How it sounds

Digital equipment is often said to sound very clean, which has to do with the 'perfect' nature of digital technology. Many musicians prefer the 'imperfect', more organic character of analog amps and effects. Still, digital equipment is gaining popularity, helped by the fact that differences in analog and digital sound are diminishing.

Why digital?

And there's more:

- Digital equipment allows for **larger ranges** in effect settings, as well as more precise settings, and you can store, recall, and edit your settings.

- Digital technology is **more controllable**, and often more affordable.

- The software can often be upgraded so you're always **up-to-date**.

- Digital technology has also made it possible to create models of classic amplifiers, effects, speaker cabinets, and microphones with **growing perfection**.

- Digital technology allows for **all kinds of extra features** on (practice) amplifiers, such as jam-along loops, beats and grooves, music players for WAV, MP3 or other file formats, a tuner, phrase trainers (see page 212), and other extras.

67

Settings

Many digital devices use traditional knobs to tweak sounds and effects. If you recall a patch, these knobs will not reflect the actual settings. Some companies have solved this problem by using motorized knobs that assume the correct positions as soon as you recall a patch. Others do the same by using knobs that are surrounded by a ring of LEDs that indicates the current position of the knob.

LEDs reflect the current settings of this control.

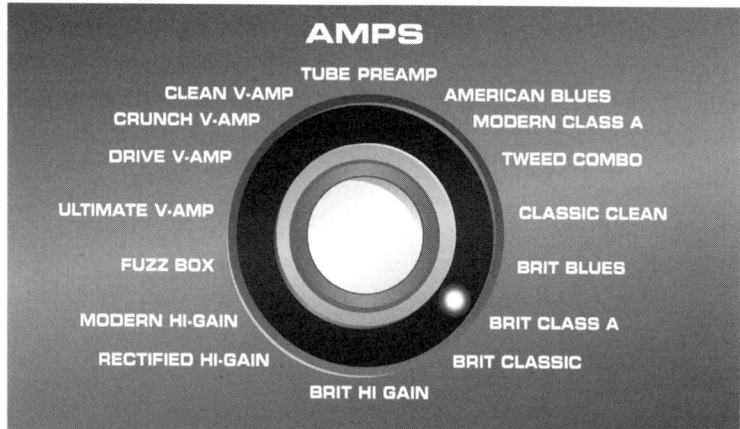

How it works

To process sound (audio), a digital unit needs to convert the audio signal to digital information, turning it into *bits*. This is done by an *analog-to-digital (A/D) converter*. The digital information is then processed by a *digital signal processor (DSP)*. Afterwards, it is turned back into an analog signal by a *D/A converter* so it can be sent to a power amp, for example.

Sampling rate

To convert an analog signal into a digital signal, it is 'measured' or *sampled* numerous times per second. The higher this number — the *sampling rate* — the better the quality of the digital sample. For audio CDs, the sampling frequency is 44.1kHz (over 44,000 samples per second).

Bit depth

The sampling rate tells you how *often* a sound is measured. The

68

bit depth or *resolution* tells you how *detailed* the sample will be. A larger bit depth (*i.e.*, more bits, or a higher resolution) allows for finer detail. CDs are recorded at 16/44.1; the bit depth is 16 bits. For digital amps and effects the numbers usually range from 16/48 to 24/96, the latter for professional equipment. These figures will be higher in the future.

Digital in and out

Digital devices often have special digital inputs and outputs (SPDIF or AES/EBU) that are especially effective in studio situations: They transmit the audio signal without any additional noise at all.

TIP

MIDI

Digital amps, effects, and instruments have one more advantage: They can communicate with each other, using a system called MIDI. These four letters stand for *Musical Instrument Digital Interface*. MIDI allows you to send messages from one unit to the other. To do so, all you need is a MIDI cable and some patience to really understand how things work.

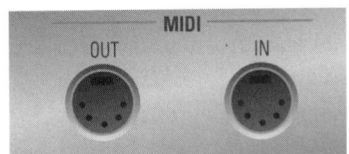

MIDI
connections.

In, out, thru

MIDI-equipped devices usually have three MIDI ports, labeled *in*, *out*, and *thru*. Here are just a few of the numerous ways to use these ports:

- The **MIDI in** port on an effects unit can be used to connect a MIDI footswitch that allows you to choose presets, switch effects on and off, change their settings, and so on.

- If you want to store all your presets in your computer (so you can edit them there, using the software that came with the

69

effects processor), you connect the processor's **MIDI out** to the MIDI in on your computer's sound card.

- **MIDI thru** ports transmit an exact copy of the information that was received at the MIDI in port to another device in a MIDI chain. An example: using the thru port, you can link a MIDI pedal board to a MIDI effects device and a MIDI amp, and operate both with the same pedal.

- MIDI can also be used to send individual presets to other devices, to exchange presets with fellow musicians, or to download new sounds, effects, or software updates, or even to control a light show — the possibilities are **endless**.

TIP

Messages, no sound

Warning: Note that MIDI ports transmit messages only, and not audio signals. Never connect MIDI ports to other types of jacks or sockets!

SPEAKERS

Speakers are to amps what drum heads are to drums: They make the air vibrate, and vibrating air is sound. Therefore, the quality and characteristics of the speaker or speakers that you're using are of great importance. When you buy a combo amp, you can't choose your own speaker(s), so you have to work with the amp/speaker combination that the manufacturer came up with. That said, you can of course always replace the built-in speakers at any time.

How?
The most visible part of a speaker is the *cone*, which is usually made of paper. This is what moves the air. The cone is attached to a *voice coil*, which can move freely within the core of the big *magnet* at the back of the speaker. When the amp sends a voltage to the voice coil, it makes the cone move. The air vibrates, and that's what you hear.

70

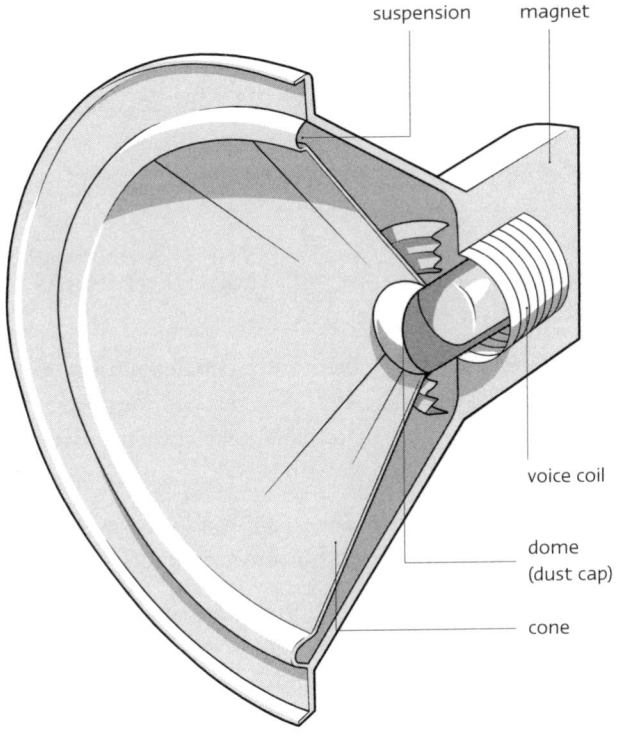

suspension   magnet

voice coil

dome
(dust cap)

cone

The flexible
suspension
(surround)
allows the
cone to move
back and
forth.

Too much

If too much signal is applied to the speaker, the cone is forced to
make a bigger move than it actually can. This produces a cracking
sound, and might actually burn the voice coil!

Large or small

Generally speaking, larger speakers move more air, making for
a larger sound and a better reproduction of low notes. Small
speakers can move faster, so they're usually better at reproducing
higher frequencies. They also offer more control and better
attack, but they're less efficient than large speakers (see page 83).
Enclosures with two or more small speakers offer speed, control,
and attack while moving as much or more air than a single large
speaker. Chapters 7 through 11 offer information on speaker sizes
and combinations for specific applications.

71

TIP

Long excursion

Bass and subwoofer speakers are typically long-excursion *speakers: They need to make large movements in order to produce the lowest notes.*

Magnets

The magnet is an important part of a speaker. Even its material plays a role, which is why it is often specified.

• Many guitarists prefer speakers with **Alnico magnets**, which were also used in vintage guitar speakers. As one of the elements for this type of material (Cobalt) became rare, these speakers have gotten quite expensive.

• Other speakers — for any kind of application — often use the more cost-effective **ferrite and ceramic magnets**.

• **Neodymium** is yet another material, featuring an equally strong magnetic field at a much lighter weight, so good speakers don't have to be very heavy anymore. With two 10" neodymium speakers, even a 300 watt bass amp can be quite portable.

Cone material

Though speaker makers have come up with various alternative materials (that may sound better or different, last longer, or both) most cones are still made of paper. Paper deteriorates with age, and cones can get damaged quite easily, so it's good to know that you can have your speakers re-coned. The suspension and other components can be replaced too.

Replacing speakers

Don't like your combo's speaker? Get a new one. Different speakers have very different sound characteristics. The main speaker makers (see page 259) offer information on the characters of each of their speakers, and some have a CD with sound samples. Still, you can't really judge what a speaker will do for you until it's installed. Apart from the sound you're after, you need to take several other things into account, ranging from the speaker's

power handling (see page 80) to the size of your enclosure, so ask for professional guidance.

Tipcode AMPS-007
Play this Tipcode to hear the sounds of five different speakers in the same combo amp.

Tweeters

Tweeters come in all shapes and sizes, often attached to a horn that helps disperse their focused high frequencies.
Piezo tweeters are commonly said not to perform too well at really high volume levels, producing a less dynamic, thinner sound than *cone tweeters* and other *dynamic tweeters*. *Slot tweeters* have a vertical slot with a metal lip, replacing the traditional horn to disperse the sound.

Too many

There are more types of tweeters than you will be able to remember. Don't pay attention to their often impressive and confusing names, but listen to what they do.

Degrees

A horn's dispersion pattern is expressed in degrees. A 60°x40° horn has a narrower dispersion than an 80°x60° horn, but the first one will probably have better projection. Some horns are designed to have equal horizontal and vertical dispersion, e.g., 65°x65°.

Dual cone

In a *dual cone* speaker, the high frequencies are taken care of by a *whizzer cone* — a tweeter that is centered in the woofer. This space-saving design is also known as a *(dual) concentric, twin cone,* or *coax(ial) speaker.*

73

PROTECTION

Amps and many effect devices have one or more *fuses* that blow in the case of an overload or a short, reducing the risk of further damage to the equipment or yourself. Fuses that are located on the outside of the amp, usually under a plastic cap on the rear side, can be easily and cheaply replaced — so always have a spare fuse of the required rating with you. (This rating or value is printed both near the fuse holder and in the manual.) Equipment that has the fuse on the inside should be taken to a technician.

More fuses
Powerful amps may also have a *heat (HT) fuse* that blows if you drive the amp too hard for too long. *Thermal protection circuits* shut down the amp when it gets too hot.

Resettable
Tweeters and other items can be protected by a *resettable circuit*: Rather than replacing a fuse, you just press a button if the circuit gets overloaded.

Fans
Amps that generate a lot of heat have one or two fans, which usually switch on automatically when the temperature gets above a certain point. Some fans produce more noise than others, which is no problem on loud stages where such powerful amps are generally used. A noisy fan in a studio doesn't work, so check if you can turn it off. When you do, be careful not to overheat your equipment. Amps can reach operating temperatures of some 190°F (90°C)!

Power conditioner
A power conditioner is a separate device that you may want to check out if you gig a lot. It protects your amp and effects from voltage overloads (which can fry your equipment); it lowers AC line noise; and it may improve your amp's sound, adding bite, attack, clarity, and punch, as well as increasing dynamics — though some experts deny all those benefits. A power conditioner

74

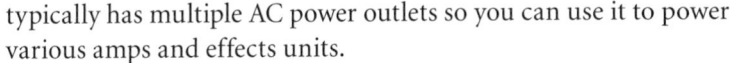

typically has multiple AC power outlets so you can use it to power various amps and effects units.

THE INSIDE

Other than what you can see when taking a look inside an open-back guitar combo, there's not much to judge when it comes to an amp's innards. Still, there are two 'inside' topics you often read about in bass or guitar amp reviews.

Circuit board or point-to-point wiring

The large majority of amps have their electronic circuit 'printed' on one or more *circuit boards,* which allows for an efficient and consistent mass production process. In expensive handmade bass and guitar amps, however, the electronic components are usually connected by *point-to-point wiring,* a very labor-intensive process. Note that there are high-end amps with circuit boards: As long as high-end components are used, the overall quality doesn't have to suffer.

Resistors

Guitar amps are the most colorful amps there are, which explains why many guitarists favor amps that use *carbon film resistors* rather than modern *metal film resistors*: The latter are sometimes said to produce a cleaner sound, and that's not what most guitarists are after. Do note, however, that you're quite unlikely to ever hear the difference between the two, and that there are plenty of other parts that are far more important for the sound of your system.

5

Figures

*When you deal with amplifiers and effects, it really helps
to have some insight into what watts and decibels are,
what impedance is, and what 440Hz or 8k stand for. This
chapter explains the basics without sending you back to
physics class.*

When selecting equipment, you'll do better trusting your ears than the device's specifications. That said, understanding what these specs mean may very well help you understand what you're buying or using.

POWER RATING

An amp's power rating is usually considered a good indication of how loud it is. In reality, however, this is just one of the many elements that play a role.

Watts RMS

The power rating of an amp is expressed in watts. There are various types of power ratings. The only relevant one indicates an amp's *continuous average power*, often referred to as *watts RMS*.

Music and peak power

An amp with a continuous power rating of 50 watts usually has a 100 watts of *music power* and some 200 watts of *burst power* or *peak power*: This means that it can deliver 200 watts to reproduce loud peaks or spikes in the music — but it can't do so continuously. When you're comparing power ratings, make sure you don't compare the continuous power rating of one amp to the peak or music rating of another. This book and most other publications on the subject generally list continuous power ratings.

How loud?

The number of watts says much less about how loud an amp can sound than you might think. For example, a 50 watt amp can sound only very slightly louder than an otherwise similar 25 watt model. To achieve twice the sound level, you need an amp that's ten times as powerful (i.e., 250 watts!). There's more on this on pages 81–82.

Why more

If a 50 watt amp sounds only perceptibly louder, why would you prefer one to a 25 watt model?

78

- A more powerful amp has more *headroom* ('reserve'), allowing it to **comfortably handle higher volume** levels — just like powerful cars are more comfortable at higher speeds.

- Powerful amps are better at reproducing **dynamic differences**. The extra headroom buys you an undistorted reproduction of sudden peaks. Peaks draw a lot of power.

- More watts often come with **more features**: more controls, channels, inputs and outputs, and so on.

Too much

Your amp should have sufficient power, but more isn't always better.

- It's no use buying very powerful amps for **very small places**: Most amps don't sound their best at their lowest volume levels.

TIP

Tips

A few tips that may come in handy:

- **Stereo amps** have two power amps. A stereo amp of 2x100 watts produces a total of 200 watts — but that's not the same as a 200 watt amplifier. The latter will have more headroom, for example.

- Some stereo amps can be **bridged**. If you **bridge** a 2x100 watt stereo amp, it turns into a 200 watt mono amp.

- Bi-amping means using **separate amps for the low and high frequency ranges**. Such 'specialized' amps can be very efficient. The amp that drives the woofer is typically more powerful than the one that drives the tweeter (see pages 166–167). Tip: These figures are often added up, but the resulting sum total promises more power than you really get.

- The amp's **power consumption**, usually printed on the rear panel, is also stated in watts.

79

- You need enough power to keep up with the loudest musician in the band — which is different from trying to become the loudest one yourself. When you do so, the others will probably do the same, and you'll end up with more power than you need and **less money than you want**. And if you really need more power, there are various ways to amp your amp, as described on pages 59–60.

- If you're looking for the sound of overdriven power tubes, a tube amp with too much power may sound **way too loud**. Want to know more? See pages 115–116.

WATTS AND SPEAKERS

Speakers also have wattage ratings. This figure tells you *nothing* about how loud it can sound; it just specifies the amount of power it can handle. A 50 watt speaker can handle 50 watts of power continuously, and much louder peaks.

RMS and AES
A speaker's power handling is usually specified in watts RMS, as well. A growing number of companies are using watts AES instead. This is a new and very well defined standard of measuring a speaker's power handling. There's no formula to go from watts RMS to watts AES or vice versa, but specifications in watts AES and RMS typically don't differ more than an almost-negligible ten percent.

Speaker and amp
If you buy a combo amp, the speaker's power handling will typically be one and a half to two times higher than the amp's power rating. You can use those figures as a guideline when connecting (extension) speakers to an amp.

Adding them up
If you connect speakers the way they are connected in a speaker cabinet (i.e., *in series*; see pages 85–86), their power handling increases. A cabinet with four 25 watt speakers has a total handling power of 4x25=100 watts.

Tips

- If the speaker's power handling is much higher than the amp's power rating, you can easily **damage the speaker** by driving the amp too hard for too long: If an amp has to work too hard to drive a speaker, the speaker may not survive. Actually, most speakers are damaged by amps that don't have enough power for the situation they're used in.

- There's no problem in connecting a **100 watt-speaker to a 200 watt-amp**, as long as you don't open up the volume control too much.

How loud?

So a speaker's wattage doesn't tell you how loud it can sound. Its *efficiency* does, and this requires some knowledge of *decibels*.

DECIBELS

The loudness of a sound — the *sound pressure level* or *audio level* — is stated in decibels (dB). A decibel can be loosely defined as the very smallest change in audio level humans can hear. If the audio level goes up 10dB, the sound is commonly defined as twice

Tipcode AMPS-008
This Tipcode demonstrates a 3dB increase in volume, requiring twice the power from the amp.

TIPCODE

81

as loud. As explained on page 78, a 10dB increase requires no less than ten times as much energy.

Double the energy

Contrary to inches or dollars, you can't simply add up decibels. To make something sound 3dB louder — a very slight increase — you need twice the energy. That explains why a 50 watt amp sounds only 3dB louder than a model with half that power.

Acoustic guitars

Another example: One acoustic guitar produces about 80dB. Two of them have twice the energy, so they produce 83dB. Four guitars produce 86dB, sixteen guitars 92dB, and so on. If you want acoustic guitars to sound as loud as a rock band, you'll need 1,224 of them!

Less and less

In reality, you can't produce 110dB no matter how many acoustic guitarists are playing, because you won't be able to hear the ones

Sound pressure levels, measured at 3.3ft/1m unless specified otherwise.

instruments	dB (at 1 meter)	sound
	00	absolute silence
	10-15	whispering
	20	leaves in the wind
	40	waves along the shore
	60	conversation
	70	shouting
acoustic guitar	80	vacuum cleaner, truck driving by
cello	83	
double bass	86	
flute	90	turning your home stereo way up
trumpet	95	
drum set	102	
	110	electric drill at 3 feet, car horn
rock band	110 –120	
	120	threshold of pain, jet plane taking of at 230ft/70m
	135	The same jet plane at 65ft/20m

sitting further away. Each time you double the distance between your ears and the sound source, the sound pressure decreases by 6dB. That explains why there are PA systems with some 200,000 or more watts of power: The people in the far back need to hear the music too.

Less power, more sound?
Speakers convert just a very small part of the energy they get into sound; the rest, some 95%, is converted into heat. The more sound you get from a speaker, the more *efficient* it is. A 35 watt combo amp with a very efficient speaker can sound noticeably louder than a 70 watt amp with a less-efficient driver.

Speaker efficiency
As mentioned earlier, speaker efficiency is expressed in decibels. Values generally range from 90 to 100dB 1watt/1m (i.e., using 1 watt, measured at 1 meter). Smaller speakers are usually less efficient than larger models.

Sound pressure level
A speaker's maximum sound pressure level or *SPL* is measured in dB as well. Powerful systems can produce as much as 135dB or more measured at 1 meter.

Your ears
Exposure to high sound pressure levels (SPL) can easily damage your hearing, either right away or in the long run. Wearing proper hearing protection can help save you from tinnitus (a perpetual ringing in the ears) or loss of hearing. To give you an idea: The US Government's Occupational Safety and Health Administration has specified that 110dB is permissible for only half an hour per day, and many bands easily produce such sound levels, even in rehearsal.

Hearing protection
When you play in a band — or when you practice on your drum set — you easily get to those high SPLs. You can protect your ears with anything from foam plastic earplugs to custom-made plugs with replaceable or adjustable filters, the latter being more expensive as well as more effective. Besides, good plugs just

83

reduce the sound level, while cheap plugs make things sound as if you're in the other room. Most music stores sell various types of affordable plugs. *Tip:* Always carry your ear plugs. If you forget them, even bits of napkins or tissue in your ears can help protect your hearing.

SPEAKER IMPEDANCE

A speaker has a certain resistance or *impedance*. The higher this resistance, the harder an amp has to work to drive the speaker. Knowing a bit about this is mainly important if you want to use extra speakers and extension cabinets.

Ohms
Impedance is rated in ohms (Ω). Most speakers and speaker combinations have an impedance of 2, 4, 8, or 16 ohms.

Lower
If you connect speakers, their impedance changes. For example, if you connect a 16-ohm speaker cabinet to a combo amp with a built-in 8-ohm speaker, their combined impedance will be 5.33 ohms*. That's good news: solid-state amps typically respond to this lowered impedance by supplying you with more power!

Second speaker, more power
If a 100 watt solid-state amp has a built-in 8-ohm speaker, connecting an 8-ohm extension cab will lower the impedance to 4 ohms. As a result, the amp will now provide you with some 150 watts of power, or even more!

Flexible
The above makes for a very flexible system: For small gigs you bring your combo amp only; for larger gigs, you get both more power and a wider sound stage by adding an extension cabinet.

84

* According to the equation $(16*8) \div (16+8) = 128 \div 24 = 5.33$.

Not too low

Unfortunately, every amp needs a minimum speaker impedance. If it gets too low, the amp will overheat as it's trying to put out more power than it is capable of. So if the bass amp from the previous paragraph had a 4-ohm speaker built in, it would have more power from the start — but adding a second speaker would lower the impedance so much that this might damage the amplifier.

The right impedance

With solid-state amps, the minimum impedance for extension cabinets is always specified. Tube amps allow you to connect cabinets with various impedances by supplying you with various speaker outputs (e.g., 16Ω, 8Ω, 4Ω, and/or 2Ω), or with an impedance switch. As long as you closely follow the instructions supplied in your amp's and cabinet's manuals, the risk of mismatching impedances is minimized. Seek advice if you're not sure.

In parallel

When you plug an extension speaker cabinet into a combo amp, the speakers will be connected *in parallel*. This decreases their total impedance according to the equation in the footnote on page 84.

In series

Within a speaker cabinet, speakers are often wired *in series*. Wiring speakers in series increases their total impedance. Two 8-ohm speakers in series have an impedance of 8+8=16 ohms.

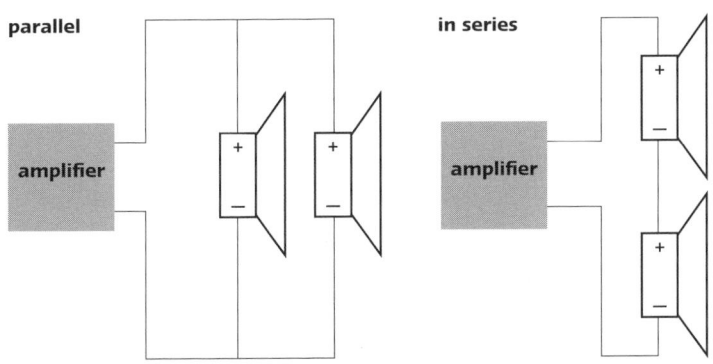

parallel   in series

amplifier   amplifier

Speakers wired in parallel and in series. As long as you're not going to wire speakers yourself, you can forget about this.

85

Combined series/parallel wirings are common in cabs with four speakers.

MORE ON DECIBELS

Besides sound pressure levels, the decibel is also used to measure other quantities, such as the *signal-to-noise* ratio and the *line input* and *line output levels* of amps, effects and other devices. The range of equalizer controls is expressed in decibels too (see page 48).

Line levels
If you connect the line output of an effects unit to the line input of an amp, their levels should match. For this purpose, the industry has agreed to a standard line level. Two, actually.

-10dB
The most common line level is -10dB. Most effects units, and keyboards, synths, CD players and other home stereo and (semi-) professional devices use this level.

+4dB
Professional equipment, from rack devices to mixing boards and PA systems, uses the higher +4dB level. (This is also notated as dBm, dBv, or dBu.)

TIP

Signal-to-noise ratio
Each amp produces a certain amount of hiss or noise. Just turn up the volume up a bit, don't play, and listen. If an amp with a -90dB signal-to-noise ratio produces a sound level of 100dB, you will hear 10dB of noise when you cut the signal (100-90=10). For instrument amps, a -90dB signal-to-noise ratio is considered very quiet.

Both

Many devices have jacks for both levels, or a +4/-10 switch. Choosing the wrong jack or level may cause distortion or a weak signal (and hiss as you're trying to boost it).

MORE ON IMPEDANCE

Just like speakers have a certain impedance, so do effects units, microphones, electric guitars, etc. The main thing you should know is that there are *low-impedance*, *high-impedance*, and *ultrahigh-impedance* devices. Instead of the word 'impedance', the letter 'Z' is often used.

High and low

Most instruments are high impedance, so instrument inputs are high impedance too. Most microphones are low impedance, so they need low-impedance or *lo-Z* inputs.

XLR is low; ¼" is high

Low-impedance inputs usually have XLR sockets (e.g., microphone cables). A ¼" jack usually indicates that you're dealing with a high-impedance or hi-Z input (such as an instrument input).

Ultrahigh, high, and low

To help prevent feedback, upright bassists and other acoustic musicians often use a *pickup* instead of a microphone (see pages 133–134 and 225–227). A pickup is a small sensor that is attached to or built into the instrument. It literally picks up vibrations and sends them to the amplifier. These pickups typically have an impedance that runs into the millions of ohms. This is usually labeled as an *ultrahigh* impedance, relative to the low impedance of professional microphones (some 150–600 ohms) and the high impedance of most instruments (thousands of ohms or *kilo ohms*). In a list:

- Pickups (e.g., piezo pickups): millions of ohms or meg(a) ohms; **ultrahigh** impedance.

87

- Electric basses or guitars: thousand of ohms (kilo ohms); **high** impedance.

- Professional microphones: hundreds of ohms; **low** impedance.

Higher to lower

If you feed a higher-impedance source into a lower-impedance input, the impedance mismatch may attenuate the signal by 10dB or more, and the sound may distort and lose brightness and dynamics. Compensating for the loss of signal by turning up the volume control will also boost hum and hiss, so it's better to use a preamp or a DI box: Both also convert the impedance.

DI boxes and preamps

To connect an electric guitar or bass to a balanced, low-impedance input (on a mixer, for example), a DI box is the most popular solution. To connect an ultrahigh-impedance pickup to a lower-impedance input, most players use a preamp.

TIP

> ## Fridges and neon lights
>
> Ultrahigh-impedance inputs, found on some bass amps and acoustic amps, can be used for lower-impedance sources such as an electric guitar or bass. However, such connections are more sensitive to noise generating devices such as neon lights, refrigerators, power cables, and power transformers.

FREQUENCIES AND HARMONICS

To understand a bit more about equalizers, speakers, the difference between tube amps and solid-state amps, and many other subjects, it's good to know a bit about frequencies and harmonics.

Vibrating air

Sound is vibrating air. If a guitarist plays the note A4 (first string, fifth position), that string vibrates 440 times per second. The speaker that reproduces that note also vibrates at that speed. *Vibrations* or *cycles per second* are stated in *hertz* or *Hz*.

Harmonics
As that string vibrates at 440Hz, it also produces numerous higher frequencies, known as harmonics or overtones. The first harmonic equals the fundamental (440Hz, in this case). The second is at 880Hz*: twice as fast, sounding one octave higher. The third harmonic is at 1320Hz, three times as fast — and so on, up to some twenty harmonics. The relative volumes of these overtones give a tone its character: They allow you to hear the difference between notes of the same pitch on a cello, a guitar, or saxophone.

Even and uneven
One of the reasons that tube amps produce a pleasing type of distortion is that overdriven tubes add *even-order harmonics* (the second, fourth, and so on) to the sound. These harmonics happen to sound more 'harmonic' than most of the odd-order harmonics that are generated by overdriven solid-state amps.

From 20 to 20,000
The human ear can hear pitches from as low as 20Hz to as high as 20,000Hz (20kHz or 20k), the latter figure getting lower as you age. When that happens, sounds lose their brightness, which is somewhat similar to what happens if you use an equalizer to roll off high frequencies.

Tipcode AMPS-009
This Tipcode plays the entire range of a piano, from the lowest key (A0; 27.5 Hz) to the highest (C8; 4,224 Hz).

TIPCODE

---

* *Strange as it may seem, the second harmonic equals the first overtone, the third harmonic equals the second overtone, and so on.*

89

From 33 to 4,224

The diagram below shows the frequency ranges of some popular instruments. The lowest C on a piano is 33Hz; the highest C is 4,224Hz. Contrary to low C, which produces many audible harmonics, high C has a very limited harmonic content: You can only hear the first four harmonics, as the fifth (5x4,224=21,200Hz) exceeds the human hearing range.

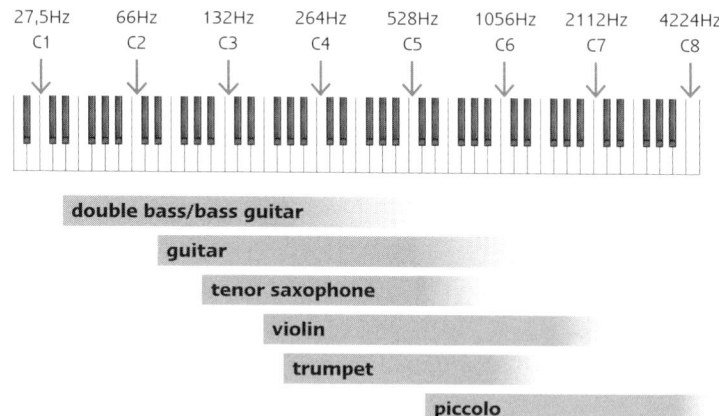

6

Testing Amps:
General Tips

The information in the previous two chapters tells you what to look for in an amp, and how to judge various facts and figures. But when it comes to selecting amps, your ears have an important job as well. This chapter covers general testing tips for any type of amp. Many of them are relevant for effects and microphones as well. More specific tips can be found in later chapters.

First of all, judging and testing amps is mainly a matter of comparing them. An amp may sound impressive at first and turn out to be rather bland, harsh, or cold when compared to another model.

Neutral?
Even 'neutral' sounding amps such as keyboard amplifiers have a character of their own. The variety among guitar amps is much larger. Good salespeople can help you make a choice based on the type of sound you're after, the music you play, and your budget.

Side by side
Don't listen to all amps that meet those conditions at once, but compare no more than two or three amps side by side. Replace the one you like the least with another model, and so on.

Your own instrument
If possible, bring your own instrument, especially if you're a guitarist or a bassist: Your amp is the other half of your sound. That said, keyboard amps might also sound better with one instrument than with another. If you use a microphone onstage, bring it with you too.

A good instrument
Of course, an amp will never sound better than the instrument you use. Most experts will tell you to get a good instrument first, and then invest in a high-quality amp, rather than the other way around: A great amp may instantly reveal the lesser qualities of your instrument.

TIP

Critical
Amps that are better at reproducing the differences between individual instruments will usually respond to subtleties in your playing too, including your dynamics and every intended (or unintended) nuance. A less critical amp may be more comfortable at first, but boring in the long run.

92

Personality

Amps strongly differ in how well they reflect the personality of the instrument you use it for. Some make every guitar or every keyboard sound alike; others stress the specific character of each instrument.

From a distance

If you sit or stand very close to a floor-mounted combo or speaker cabinet, you won't be able to really judge its sound. Direct the speakers at your ears, or better yet, listen from a distance. Have someone else play it, and pretend you're the audience: In the end, you'll be playing for their ears.

Carrying the sound

Also have someone else play at a moderate volume and walk around the store to see how well the amp and speakers carry the sound across the room. Some combos and speaker cabs seem to lose a lot of sound at no more than fifteen feet, or the sound gets harsh or boomy as the distance increases. *Tip:* Room acoustics do play a role here too!

Sound dispersion

The sound should be focused, rather than washing out. Don't confuse this with the speakers' ability to spread the sound around. For most applications, the speakers (especially the tweeters) should have a wide enough *sound dispersion pattern*. Have someone else play or hook up a CD player, and see if the sound changes when you move off axis, away from the speaker's line of fire: Some speakers are more directional than others.

Equalizer

When checking acoustic guitar amps, bass amps, or keyboard amps it's usually best to set all tone controls in their flat positions. If you don't, you're comparing equalizers rather than the amps themselves. Start fooling around with EQ settings only once you've narrowed down your choice to two or three amps. When trying out the EQ, listen how the tone controls influence the various frequency ranges, and judge their range. Some EQs are more effective than others. *Tip:* Most electric guitarists use their EQ in a com-

93

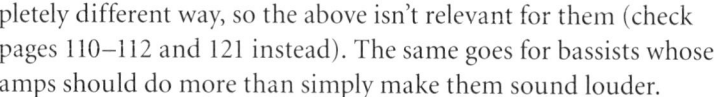

pletely different way, so the above isn't relevant for them (check pages 110–112 and 121 instead). The same goes for bassists whose amps should do more than simply make them sound louder.

Power

An amp may sound great at relatively low settings and much less impressive when cranked past 5. Others only shine at higher volumes. Likewise, there are amps that sound very loud but turn out not to be able to cut through the sound of your band. These things can't be judged in the average store, which again stresses the importance of a return policy.

What?

So what can you listen for in a store? Listen to the sound as a whole, but also listen to what happens in the three main ranges: bass, mid and treble.

- **The bass range** can be well controlled, tight, fat, and big, or boomy, wooly, and muddy...

- **Midrange frequencies** provide detail and intelligibility, but they can be aggressive, harsh, and ear fatiguing, rather than accurate, warm and natural...

- **High frequencies** provide crispness, clarity, and brilliance to the sound, but they can also be shrill, brittle, or piercing...

Bright or shrill

What one finds bright (and attractive) another may perceive as shrill (and unattractive). Still, using such subjective terms can make it easier to find what you're looking for.

Just as loud

If you do a side-by-side comparison, make sure all amps sound equally loud. If not, you're bound to prefer the one that sounds loudest, even if you can hardly hear the difference in volume. Boosting the bass on one of the amps typically has the same effect.

Noise

Listen for noise, varying from low hums to high-frequency hiss. Don't play, but just turn up the volume and listen. Guitar amps

usually won't be as quiet as other amps can be, especially when using the lead channel: They always hiss. On high-powered amps, a fan may disrupt the silence, but this is only a problem if you can hear it onstage as well.

Standing waves

A boomy type of sound may indicate standing waves, a result of acoustic reflections in the speaker enclosure or in the room. Simply altering the enclosure's position will tell you which one it is. Many enclosures have an asymmetrical (internal and/or external) design to prevent standing waves.

Breaking in speakers
Do speakers need a break-in period, loosening them up a little so they respond faster and produce more body and low end? Experts disagree, and most of them seem to say no. That said, an original vintage speaker that has been used for dozens of years will never sound the same as a brand new replica of that same speaker.

Preamp only
If you can hook the amp up to a PA or a recording console, those devices will receive the preamp signal only. Check out what that sounds like, if you plan to use your amp that way: A combo amp that sounds pretty good as a whole may still have a disappointing preamp. Electric guitarists should check if the line out has speaker emulation (see page 57).

Radio?
Both affordable and professional amps may pick up radio signals. Note that a poor-quality instrument cable may be the culprit too, as discussed in Chapter 15.

Headphones
Some amps sound very different when you use headphones. If you do so a lot, you may need to change your settings every time you switch from using the speakers to using headphones.

95

SECONDHAND

If you're buying secondhand equipment, there are a few more things to pay attention to.

- Check **all of the controls**. Connect your instrument, turn the volume up a bit and rotate the knobs to see if they're working smoothly and properly.

- **Scratchy pots** can usually be fixed easily (see page 246).

- Listen for noise while plugging your instrument cable **in and out**.

- **Rotating the plug** should not produce dropouts or noise.

- Play and make sure **the speaker** doesn't produce more sounds than it should.

- To really check out a speaker, gently **push the cone** in and let it come back. This should be a noiseless operation. Note that you may damage a speaker this way — so be careful.

- Can you **store presets**? Then edit a sound, store it, and recall it to see if it all works.

- **Listen**. Amps get audibly older, losing power and body. Restoring an old amp with new components (resistors, capacitors, etc.) will change its sound!

- Buying a tube amp? Check **the tubes** (see page 248).

- Watch out for **modified amps**: Modifications may bring about safety hazards.

- **A manual** is handy for many amps, and indispensable for most programmable models with multi-function controls.

7

Guitar Amps

Unlike most other amps, guitar amplifiers are designed to color the guitar's sound, rather than to just amplify the input. That explains why many guitarists consider their amp an instrument in its own right — one that you need to learn to play to make it sound the way you want.

A guitar amp can be anything from a 2 watt practice amp to a 300 watt monster, from a basic unit with a single channel, single volume control, and two tone knobs to digital devices with dozens of virtual amp and cabinet models. This chapter covers their main differences and features.

Character

Guitar amps have very pronounced characters, and a lot of them are designed with a certain style of music in mind. Some brands are famous for their clean, controlled-sounding jazz amps; others are just as well known for their brutal, in-your-face distortion, and still others prefer making versatile amps that can be used in a wide variety of styles.

TIPCODE

Tipcode AMPS-007
Play this Tipcode to hear the sounds of five different speakers in the same combo amp.

British, American, Modern

When talking about amp characters, you'll often hear the terms *British*, *American*, and *Modern*. These primarily refer to the overdriven, lead channel sounds of these amps.

- The original **British sound** has well-defined, aggressive, gritty mids that give the notes plenty of punch and a meaty or even raunchy tone with lots of sustain. However, the modern British lead sound is quite different, with the mid control set to scoop out the midrange response.

- The **American sound** has a fuller, looser bottom end and a denser, more rounded tone.

- The term **Modern** usually indicates an American-based, higher-gain version of the modern British sound: heavily distorted,

98

lots of gain, a fat low end and scooped mids. Think fat, bad, and metal.

Tipcode AMPS-010
This Tipcode offers brief examples of three characteristic guitar sounds: British, American, and 'modern'.

TIPCODE

Confusing

With subjective terms like these, you may very well run into other descriptions. Also, there are US-made amps that sound very 'British', as well as British amps with an 'American' tone.

Pre or post distortion

A meaningful difference is that American amps typically have a pre-distortion EQ, while many British amps have their tone controls post distortion.

More or less effect

As distortion adds a lot of harmonic frequencies, a post-distortion EQ commonly has a much more dramatic effect: The tone controls affect the distorted sound, so there's simply more material to cut or boost. This works best for high-gain distortion. A pre-

Listen up

Amplifiers are straightforward devices that simply follow basic physical laws, or so you'd say. Still, experts don't always agree on some of the subjects that are covered in this book. For example, some say that a post-distortion equalizer is less effective. The truth? Go out, listen to a variety of amps and settings, and see what you think.

TIP

99

distortion EQ typically yields smoother results, allowing you to limit the frequencies that you are going to distort. *Tip:* Some amps offer both pre- and post-distortion EQ.

TUBE OR SOLID STATE?

When you're buying a guitar amp, one of the main questions will be whether you're getting a tube amp, a solid-state amp, or a combination of both. Here are some practical and musical considerations.

Price
Tube amps are more expensive than solid-state amps, typically starting around six hundred dollars for a good, small amp with a minimum of features. The fact that the tubes need to be replaced every one to three years (see page 248) adds to their expense.

On the road
Tube amps are a lot heavier and more vulnerable than solid-state amps. That explains why some guitarists get a solid-state amp, even though they may prefer the sound of a tube amp.

Blindfold test
To get a clear picture of what the differences really are, it's best to sit down and take your time to compare a good solid-state amp and an equally good tube amp. Turn away from the amps or get a blindfold. You may be surprised to discover that it can be quite hard to tell the difference.

Soft clipping
Tube amps are typically preferred because of their warmer, richer, more organic type of sound, both clean and when overdriven. Overdriving a solid-state amp will literally 'clip' the signal, squaring it off. This is known as *hard clipping*. An overdriven tube amp rounds the signal off — also known as *soft clipping* — which sounds much less obtrusive.

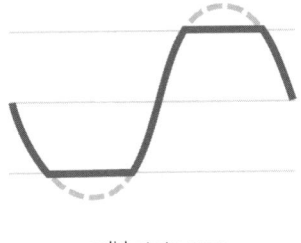

solid-state amp

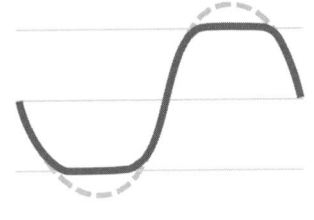

tube amp

Overdriven solid-state amps clip the signal, squaring it off. Tube amps round the signal off.

Harmonics

When overdriving a tube amp, the tubes add even-order harmonics to the tone, making it sound warmer, fuller, and fatter. Overdriven solid-state amps predominantly add odd-order harmonics. Note that this harsher type of distortion is the preferred type of sound for some styles of music, e.g., trash metal.

Compression

An overdriven tube amp compresses the sound, reducing volume level differences. If you use this characteristic well, you can go from a clean sound to a distorted sound by simply hitting your strings harder or turning up your guitar volume control: The amp's compression won't make you sound (much) louder. In this sense, tube amps are more responsive to your touch. Solid-state amps, which have less natural compression, are more *dynamic*, in the literal sense of the word: A heavier touch will simply make you sound louder.

TIP

> ### Input level
> *Tube amps respond differently to varying the input level. See what happens if you turn down the volume control on your guitar a bit. Is there less distortion right away, or don't you hear that much of a difference? Try the same with other amps.*

Louder

Tube amps are often perceived to sound louder than solid-state amps, and they are indeed. How come? The impedance of a speaker

101

depends on the frequencies it's reproducing. An 8-ohm speaker will have an impedance of 8 ohms at 400Hz, but it can be 16, 32 or even more ohms at other frequencies, as shown in the illustration on page 113. This increase in resistance strongly reduces the available power of a solid-state amp at those frequencies. A tube amp, on the contrary, continuously provides 80 to 90% of its power, regardless of the frequencies your guitar produces.

Subjectively louder

Also, tube amps tend to sound 'subjectively louder' because of the way they distort, among other things. And because their distortion tends to be more agreeable to the ear, you can play them even a little louder that you would solid-state amps, some say.

Damping factor

Another difference in tube and solid-state amps is their *damping factor*. This is the ability of the amp to control the woofer after it has set it in motion. Most tube amps are less effective at this, generating a looser, more open type of sound, while solid-state amps usually have more control, punch and definition. Some amps have a variable damping or *resonance control*, allowing you to make the amp sound loose, deep and open, or a bit tighter.

A three-channel tube preamp with speaker simulation. (Hughes & Kettner)

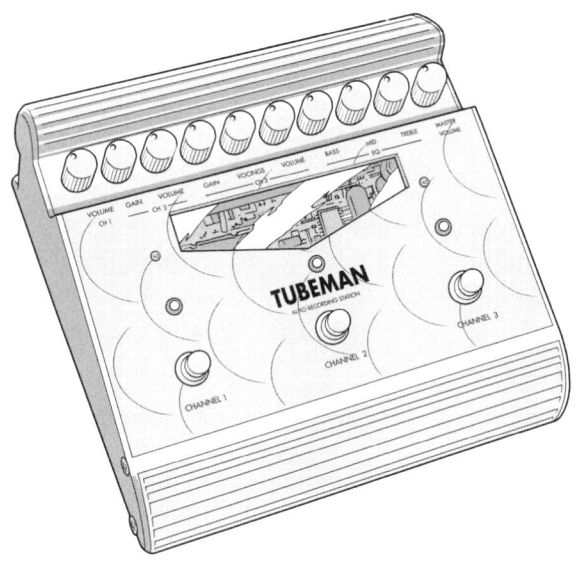

Hybrid amps

You can also check out amps that have both tube and solid-state circuitry. These hybrid amps usually have a tube preamp that 'warms up' the signal, and a solid-state power amp. This typically makes them sound closer to solid-state amps than to tube amps.

Tube preamp

Do you have a solid-state amp and are you looking for the added warmth of a tube? Then check out some of the many tube preamps available (see pages 102 and 182). If your amp has a power amp input or an effect return jack, connecting a tube preamp to it will turn the amp into a hybrid amp. Preamps that feature speaker simulation can also be used for recording or as a 'spare' amp for performances in large venues: Simply hook it up to the board.

Rectifier

The *rectifier* is the part that converts the AC electricity from the power outlet to the DC electricity that your amp runs on — and even that part influences the sound. A few amps offer switchable tube and solid-state rectifiers. Select the first for a softer, bluesy feel, more sensitivity, and less headroom, and the latter for a more aggressive, tighter, heavy rock type of sound.

MODELING AMPS

Ever since their introduction in the mid-1990s, modeling amps

Tipcode AMPS-003
Here's a brief demonstration of four digital models of well-known classic guitar amplifiers, all produced by the same modeling amp.

TIPCODE

103

have become increasingly popular, with prices of entry-level models plummeting over the years. Modeling amps allow you to choose from a number of amp models and most let you combine these with a variety of cab models. Effects or effect models are built-in as well. A growing number of modeling amps — with or without a built-in tube preamp — even allow you to create your own amps or mix the characters of various models into a new one.

Differences

It has become harder and harder to hear the differences between virtual amp models and the original amps. Still, there will always be musicians who prefer to use the real thing, often stating they'd rather have one great sound than a thousand sounds that come close.

Names

Modeling amps commonly use fantasy names to refer to popular amps from the past and the present, 'Rectified Hi Gain' indicating a Mesa/Boogie Dual Rectifier, and 'British Classic' for a Marshall Plexi 100 watt amp, for example. The names of the originals are always listed in the manual. Other modeling amps aren't going for a perfect imitation of classic amplifiers; instead, they provide you

Modeling amp. (Line 6)

Amp selector

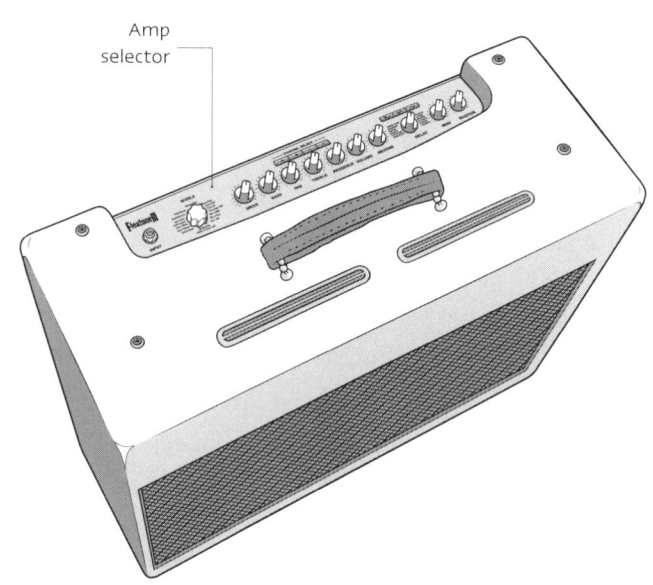

with certain types of sounds, ranging from 'Heavy Crunch' to 'Hi-Gain' or 'Jazz.'

Controls

On advanced modeling amps, the control options are also modeled after the original. If you select a vintage amp with a single tone knob, the other tone controls will be deactivated or assigned to other functions. Some modeling amps even adapt their output power rating to match the rating of the original: If you select a 15 watt vintage amp, the modeling amp will supply you with 15 watts of power, and no more.

Cabinets

Besides sporting between a handful and thirty or more amp models, modeling amps also have a variety of virtual speaker cabinets on board, so you can mix and match amps and cabs to find your favorite combination. Some amps even allow further variations, letting you turn open-backed cabinets into closed-back models (see pages 114–115), for example.

Effects

Modeling amps usually have digital effects built in, and some feature a number of stompbox models too. Special features can also include pickup modeling (so you can make single-coils sound like humbuckers, or vice versa) and acoustic guitar simulations. The same goes for modeling *pre*amps, which are dealt with in Chapter 13.

Store your settings

Modeling amps use digital technology, which allows you to store

Computer editing

Editing your presets can be a lot easier on your computer, so many companies include software for this purpose. You connect the modeling amp to your computer using MIDI or a USB connection. Always check if the software is compatible with your operating system and the computer you're using.

105

your settings (choice of amp model, speaker, and effect combinations and settings) in a number of *user presets* or *patches*. Most amps also have a number of preprogrammed *factory presets* that you can use as starting points. The number of presets varies from just a few to a hundred or more. There's more on presets on pages 206–207.

DIFFERENT AMPS

Occasionally, guitarists use non-guitar amps. The Fender Bassman, originally designed for bassists, is just one example. Other bass amps are used too, either clean or with an effects unit. Modeling preamps are sometimes used with a keyboard amp, rather than a guitar amp: The preamp's built-in amp and cabinet models provide the tone you're after, and the keyboard amp does nothing but amplify it, just like a PA.

POWER RATING

For many bands and performances, a 75 to 100 watt solid-state amp or a 40 to 60 watt tube amp with an efficient 12" speaker will do the job. You can get more than that, but less may be sufficient too. As stated in Chapter 5, a 25 watt amp is just a little quieter than the same amp with twice the power. *Tip:* Play and listen first, and then check out the amp's power rating.

High-powered amps
You can also get too much power. If you're looking for that 'sweet, saturated tube sound' and overdrive the power section of an 100 watt tube amp for that purpose (see page 115), it'll generate way too much sound for all but the largest venues. You can do something about this, though, as covered later in that same area of the book. Do you really want a lot of power? There are tube amps rated at 180 watts, and solid-state amps that supply you with 300

106

watts — which should be plenty. If there's a lot of low end in your guitar sound, if you use a lower than standard tuning, or if you play a seven-string guitar, you may check out high-powered amps. Low frequencies need more power.

> **A small amp**
> Some points in favor of lighter amps? They're easier to set up on small stages, they're less expensive, and easier to handle. If you need more power, you can amplify your amp through the PA, using a microphone or a line out or DI, for example.

TIP

Can they hear me?
Whether everyone can hear you depends on many more things than the amp's power rating, including the amp's design, the speaker's efficiency, etc.

GUITAR AMP CHANNELS AND MODES

Both low- and high-budget amps may come with one channel only. Low-budget solid-state amps often have just one channel to save costs. Some have a *mode switch* for an overdrive type of sound, the switch producing a change in level or EQ settings, which provides an instant variation to a channel. Mode switches are used on multi-channel amps too.

Distortion pedal
If you don't have a lot of money to spare, you may want to get a single-channel amp with an agreeable clean sound and add a distortion pedal to it, using that effect as your 'second channel.'

One channel, high budget
Some high-budget (boutique) amps have just one channel too. These purist amps are supposed to be played with their volume

107

control wide open, switching from a clean sound to a distorted sound using your guitar's volume control and your touch. How well the amp responds to these variations is referred to as its *touch sensitivity*.

Modeling amps
Modeling amps usually have a single channel as well. Rather than switching channels for a different sound, you just use a different model.

Two channels
On two-channel amps, the second one is usually the overdrive or lead channel. Some amps have mode switches for both channels, providing you with a choice between either a clean or a mild overdrive sound on the first, and either a vintage overdrive or an aggressive high-gain overdrive on the second channel. To switch channels, most amps come with a footswitch (see pages 42–43).

Control panel of a guitar amplifier with three chanels.

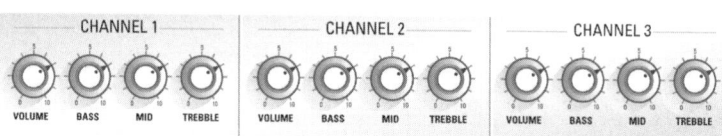

More channels
On most multi-channel amps, each consecutive channel is capable of higher gain levels. For example, the channels on a four-channel amp may be labeled clean, drive or crunch, overdrive or high gain, and ultra gain respectively. (Other multi-channel amps have two different clean tones.) Some amps allow you to mix the tonal characters of the channels for even more variation.

Comparing channels
If you listen to the clean channels of various amps, you will hear major differences, but nothing compared to what you will hear when you compare their overdrive or high-gain channels: After all, the more colors or frequencies there are, the larger the diversity will be. The trick is, of course, to find the amp that has both the clean and the dirty sound you love.

108

Two

Should you prefer the clean sound of one amp, and the overdrive of another, you can always consider buying both amps, using a special footswitch to go from one to the other. Such an *A/B box* has a single instrument input and two outputs, A and B, one for each amp.

INPUTS AND CONTROLS

Multiple channels often come with multiple controls, or with multiple inputs.

Two inputs

As mentioned in Chapter 2, guitar amps often have two inputs: one marked high, passive or 0dB, and another one labeled low, active or −6dB (the exact figure may vary). The 'low' input is less sensitive and meant to prevent undesired distortion by guitars with a high output signal or active pickups: Low stands for low gain, and -6dB indicates a 6dB attenuation of the signal. Most guitarists always use the first, 'regular,' input, which explains why many amps just have that one. *Tip:* Using the less sensitive input makes guitars with 'regular' passive pickups produce a darker tone.

Two guitars

Though they're not intended for this purpose, you can use those two inputs to hook up two guitars, for example for a practice session with a fellow guitarist. Of course, the instruments will share the amp's settings, but that usually won't be a problem under those circumstances. This works best if the instruments have similar output levels. If not, one guitar may sound noticeably louder than the other. Also, the instruments may not sound their best when amplified this way.

Inputs per channel

If an amp has one or two separate inputs for each channel, you can most definitely share it with a fellow guitarist: You can connect the

109

rhythm guitar to the clean channel and the solo guitar to the lead channel, for example.

Two channels, three volume controls

With two channels, it's good to have three volume controls: one for each channel, so you can create a good volume balance between channels, and a third one — the master volume — to change the overall level without affecting that balance. *Tip:* Some companies label the lead channel's volume control *post gain* rather than 'volume.'

Tone controls per channel

Having independent tone controls per channel allows you to shape the sound for each channel, rather than having to find a compromise setting that works well for both clean and overdrive sounds. No independent tone controls? Then consider adding a stand-alone equalizer to your rig (see page 184).

Effect level, reverb level

The more independent controls you have, the better you can tweak the amp. Some amps also have independent reverb controls for both channels, or independent effect level controls for the effects loop, for example.

GUITAR AMP EQUALIZERS

For electric guitar amps, the equalizer is a real tone-shaping device rather than something to tune your sound to the room's acoustics. Boosting and cutting frequencies can make or break your tone, and amps with a lackluster lead channel may produce a convincing lead sound once the EQ has been adjusted properly.

Independent controls

On amps with independent tone controls, the controls may affect different bands per channel. For example, the treble control might affect the 5kHz range for the clean channel, and the 2kHz range for the overdrive or lead channel.

Tipcode AMPS-004
This Tipcode is a brief demonstration of various EQ settings for jazz, blues, and metal.

EQ and gain

The tone controls can be part of the amp's gain structure. If so, turning them all down equals turning the gain control back to zero. Tone controls can affect the gain in other ways too: There are dual-purpose mid controls that act as a regular mid control on lower settings and turn into a veritable gain control when you crank them up, for example.

Switches and inputs

Next to the regular tone knobs, guitar amps often have a number of tone switches. These vary from the self-explanatory 'mid boost' to the more enigmatic 'depth' (boosts lows and low mids, rolls off highs), 'expand' (boosts lows and highs), and 'tone shift' or 'mid shift' (cuts mids, boosts lows and highs) switches. You might also see switches and knobs with names such as 'bite,' 'punch,' 'density control,' or 'attitude.' Some amps feature an additional input, labeled 'bright', providing a brighter sound than the regular input.

Presence

The function of the presence control seems to differ from amp to amp (and from expert to expert). Some describe it as a master tone control; others state that it boosts the upper mid frequency range, or that it acts as a loudness control for the very highest frequencies, adding a buzzy top end if cranked up too much. Yet another description is that you use it to fine-tune the amp's damping factor at high frequencies. And, finally, a very straightforward one: The presence control simply makes the instrument more 'present' in the mix. An amp may have independent presence controls per channel, a single control for the lead channel, or a master control for both or all channels.

111

Tipcode AMPS-011
Play this Tipcode to hear what opening up the presence control does to your sound.

Equalizers and styles

Tip: Find out the commonly used EQ settings in the style of music you play. For example, heavy metal typically requires scooped mids and strong highs and lows. Funky riffs often sound best with reduced mids and some additional treble, and for a bluesy sound you may want to cut the treble and boost the midrange a bit. Most manuals provide you with sample settings.

EQ pedals

Because tone shaping is so important, some experts recommend buying an EQ pedal as your first extra, no matter how deep your amp's EQ is. *Tip:* You can use a stand-alone equalizer for other things as well, as you can read on page 184.

GUITAR SPEAKERS

Speakers in guitar amps range from 6" to 15". Practice amps often use single speakers up to 10". Onstage, you'll usually need more or larger speakers. Combos and cabinets with one or two 12" speakers are very popular: This size is commonly considered the best speaker for the guitar's frequencies. A 15" speaker provides more warmth and fatness.

112

Different speakers

Unlike most other types of speakers, guitar speakers add a lot
of color to the sound by cutting some frequencies and boosting
others: They're the most *non-linear* speakers around. Modern
guitar speakers are even designed to mimic the imperfections of
original vintage models. A 'perfect' speaker is not what you want
in a guitar cabinet.

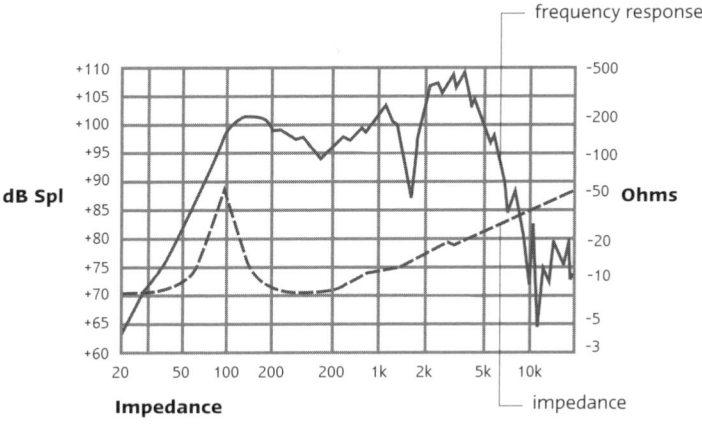

*Guitar
speakers
have a very
non-linear
frequency
response. This
diagram also
shows how
the impedance
changes along
with the
reproduced
frequencies,
up to some
50 ohms (see
page 102)!*

Character

It follows that guitar speakers have widely divergent characters,
from smooth, creamy, or warm to honky, bright, or raunchy, with
boosted or scooped mids, extended highs or fat lows, and so on.
Likewise, some speakers are at their best with heavily distorted
sounds, but they're a bit too loose for clean chords and solos;
others are great for playing clean, but they may be too 'tight' for
a convincing high-gain sound. And then there are speakers that
seem to do both equally well.

Different speakers

Some brands sell combos and cabinets with two different speakers,
aiming to create a new sound by combining their characters. Very
few companies offer a choice of speakers in combos or cabinets.

Combinations

A single 10" speaker will usually lack the power required for
live use, but using two or four of them makes up for this. For

113

example, 4x10 cabinets are very popular for an 'American' sound. Compared to a 4x10, a 2x12 will have a similar efficiency and usually a warmer, rounder timbre. Cabinets with four 12" speakers are popular among rock and metal players.

TIP

> ### Stacks
>
> Cabinets with two or more speakers are often part of a stack. In a full stack, the bottom cabinet usually has a flat front (straight or straight-face cab); the top one has an angled front (slant or slanted cab).

Stereo

There are cabinets that can be used both mono and stereo, e.g., a mono 4x12 can be used as a pair of stereo 2x12 cabs. As long as the stereo signal isn't sent to a PA system (miking both halves separately) you won't really enjoy much of a stereo effect. The same goes for stereo combos: The speakers are just too close to each other. *Tip:* Most combos with two speakers are not stereo. The second speaker may be just there for additional power handling and a wider sound image. Another tip: Some companies sell *satellite combos* or *slaves* that turn a mono combo into a stereo system with a natural stereo spread.

OPEN BACK, CLOSED BACK

Most guitar combos have an *open back*, reducing the reproduction of low frequencies and making for a more 'open' sound and a wider sound stage — especially if the sound from the back is reflected by a nearby wall. *Closed-back* enclosures, which generate more punch, lows and focus, lend themselves very well to dropped tunings and low-B guitar sounds. If this is the type of sound you're after, check out both closed-back combos and stacks: speaker cabs typically have a closed back. Low-budget combo amps often have a closed back as this is the only way to give them proper bass response and add a little extra body to the sound.

Open or closed

Some companies make combos with a removable back panel, or they offer an optional panel, so you can switch from open- to closed-back response. Of course, you may remove the back panel of any combo or cabinet, but the result will probably not be what you're looking for. Closing open-backed enclosures may result in overheating the amp, and other problems.

Careful

Open-back combos invite you to store cables, effects pedals and other accessories inside, but be careful: It takes just one wrong move for these items to damage your vulnerable speaker cones, tubes, or other parts. Some combos come have Velcro strips or other locking systems to keep things in place.

DISTORTION AND TUBE AMPS

Many players distinguish *soft* and *hard distortion*. Hard distortion, generated by overdriving the preamp, is popular for high-gain guitar sounds. Soft distortion, which has a more liquid, creamy, or 'brown' character, stems from power-tube saturation: This is the original type of distortion that was created when guitarists started cranking low-wattage amps, overdriving the power tubes.

Master volume

When overdriving the preamp, the gain knob controls the amount of overdrive, while the master volume knob helps you limit the overall volume. Conversely, if you want to overdrive the power amp, you need to crank the master volume control way up to saturate the power tubes. When you do, you'll find out that a 10 watt amp can be enough to overpower your drummer.

Triode or pentode

So how can you manage the overall volume level of an overdriven power amp if not with the master volume control? First, some amps have a *triode/pentode switch*, which switches between these

115

two power-tube modes. The pentode setting provides full power, maximum headroom, optimum articulation and hard-clip distortion should you still overdrive the amp. The triode setting reduces the amp's output power and provides soft clipping with an elastic feel and a rich, saturated sound.

Half power

Another type of switch can be simply labeled *power mode, full/ half power,* or *stage mode/studio mode,* for example: Stage mode is full power, studio mode is half power. Some amps can be switched from, say, a 100 to 75, 50 or 25 watts with a combination of triode/ pentode and half power controls, and there are amps that can be adjusted from 30 to less than 1 watt. The lower the power, the easier it is to overdrive the power amp.

Attenuator

No such switches available? Then get a *power attenuator*: an (adjustable) device that literally soaks up all or part of the power that would otherwise be sent to the speakers. Power attenuators are connected between the power amp and the speakers. They're available in rack-mountable, amp-top and other formats, and some amps have one built in. A heat sink, a fan or a vented housing help dissipate the heat they generate. Amplifier makers

A power attenuator or power brake. (Marshall)

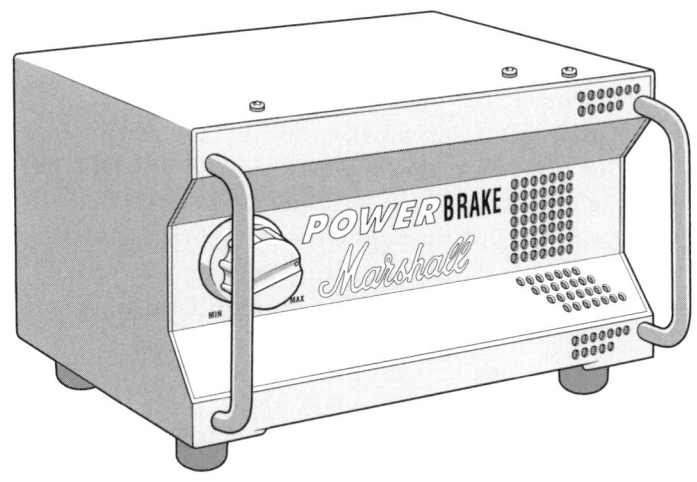

and musicians use a variety of illustrative names for power attenuators, such as power brake, load box, and power soak. *Tip:* Using a power attenuator can tempt you to drive your amp to the max all of the time. As a result, your power tubes will wear out a lot faster than they normally would.

Dummy load

Tube amps with a headphone jack need a power attenuator to act as a dummy load if you disconnect the speaker. Without a speaker or a dummy load, the amp will get damaged. There are stand-alone units that supply the dummy load and feature DI, headphone, and other outputs as well.

Class A and A/B

Most tube amps have one or more pairs of tubes that work in an alternating *push-pull* manner. They divide the workload, so to speak. This is known as *class A/B*. In a *class A* amp, all of the tubes continuously run at full power. This is less efficient than class A/B: A lot of heat is generated, and the output power is reduced by some 30 to 40%. Some amps can be switched from class A to class A/B for reasons mentioned above. Class A amps are expensive, but many players love them for their sweet harmonic distortion.

TIP

Quality?

Many people tend to think that a class A amp is simply better than a class A/B amp – but it's basically a matter of what you're looking for. If you need lots of power and headroom, you'll probably end up choosing a class A/B amplifier. A class A amp typically provides you with less power and more of a vintage type of sound.

More gain

To increase an amp's maximum gain, amp designers can use additional gain *stages*, with each extra stage providing more amplification and color. Amps with four or five stages are not uncommon.

117

TUBES

There are dozens of different types of tubes, each having a specific effect on the resulting sound. A widely used preamp tube is the 12AX7, but other types are used as well.

Power-amp tubes

There are five main types of power-amp tubes. They can be found in bass amps too.

* **6L6** tubes are an essential element of the American rock and roll and blues sound, providing warmth, detail, and a tight overdrive.

* **6V6** tubes are lighter, 'sweet and soft-clip' tubes, used in amps up to some 25 watts.

* **EL34** tubes help produce the typical 'British' rock tone with plenty of crunch.

* **EL84** tubes are used in less powerful amps. They're tighter, brighter, and more aggressive than 6V6s. Some call them crunchy.

* **6550** tubes provide more power than EL34s. They have a tight bottom end even when overdriven, and they stay clean at high volumes.

Switchable

There are a few amps that allow you to choose your own tubes, and some are even switchable (e.g., from 6V6s to EL84s). As a

TIP

Different brand, different sound

The type of tubes you use are crucial to your sound, their different characters mainly showing when you're overdriving the amp. Do note that the exact same types tubes from different brands may yield different sounds!

general rule, you should stick with the tubes that the amp was designed for. More about replacing tubes — a recurring necessity — is on pages 248–249.

BUILT-IN EFFECTS

For guitar amps, the most common built-in effect is a *spring reverb*: a metal box or *tank* with a number of rather loosely coiled springs inside. The signal is sent through these springs, which adds the ambient reverb effect to the sound. Bumping an amp with a built-in spring reverb causes a loud 'boing,' due to these same springs.

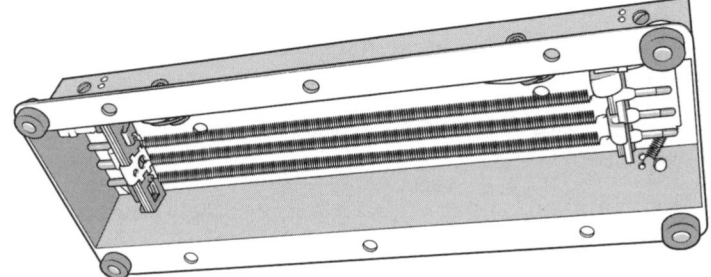

*A short (9.25")
reverb unit with
three springs.
The long
version is 16.75"
(42.5 cm).
(Accutronics)*

The springs
Spring reverbs come in various lengths, commonly with two or three springs. More and longer springs produce a richer, warmer, and deeper reverb. Tube-driven spring reverbs sound warmer than solid-state versions, but they're more vulnerable. Good spring reverbs have a transparent, non-metallic sound, and they don't cut the bass frequencies too much. In used amps, check to see if the springs are in good condition. They shouldn't be damaged or too loose.

Chorus and 'vibrato'
A chorus is much less common than a spring reverb, but some guitar amps have one built in. Others have an integrated tremolo (a rapid volume alteration). Tremolo is often incorrectly referred to as a vibrato, which is a rapid pitch alteration.

119

More effects

Most amps with a built-in multi-effects processor are modeling amps. These amps are compact, there's no hassle setting things up, and you usually pay less than you would for a separate amp and effects processor — but you can't choose your own processor.

TESTING GUITAR AMPS: TIPS

- Check the **gain and volume controls**. Does the distortion ramp up gradually, or does it come on all at once? Some amps sound amazingly loud at 1 and start crunching at 2. Does that mean it's a loud amp? Not really: It usually means that it doesn't get much louder beyond those settings, as compression sets in.

- Listen if (and how) the **tonal character** changes as you control the master volume, and do the same with the gain control(s).

- Lots of gain can make an amp sound really loud, but does **it cut** through the band's sound?

- When comparing guitar amps, try to **tweak their tone controls** so each amp sounds its best and then judge their sounds. First compare the clean channels of each amp, then the other channel(s), or the other way around.

- **Clean channels** can be crisp, sweet, deep, thick, warm, smooth, punchy, sparkling, woody, ringing, and well-rounded, or sharp, one-dimensional, brittle, muddy, thin, and so on.

- **Lead channels** can sound powerful, crisp, focused, full, creamy, bluesy, in-your-face, brutal, and rich, or edgy, thin, brittle, stale, gritty, fuzzy, indistinct, wooly, and lacking texture or focus, for example. These are subjective terms, of course. What jazz players call warm, others may call indistinct, and what country guitarists find clean may sound sterile to your ears.

- Listen to all of the channels over the amp's **entire volume range** and see if the sounds hold up. Can you still hear every note at high-gain settings, even when playing complex chords or fast licks?

- Also check out how an amp handles different **distortion levels**. Play every amp the way you intend to play it live.

- The **tone controls** are extremely important for shaping your lead sound. Take your time to see what they do and how to use them. One simple trick is to try them at their extreme settings. Start with maximum bass and treble, and minimum midrange, for example, and then slowly vary the range as you play.

- When testing and comparing amps, set your guitar's tone controls in their **neutral or flat position** (i.e., fully open on passive guitars; middle setting on active guitars).

- Auditioning tube amps? See if you can overdrive them by simply **hitting your strings harder**.

- Check the amp's **channel switching**. The sound should not cut out when switching channels. The same goes for changing amp models when you're using a modeling amp.

- Try amps **without any effects** first, and add your own or the onboard effects once you've narrowed your choices.

- Listen for **noise and hum**, especially when using the lead channel. Turn the volume up, and don't play.

- Some amps work great with **certain types of guitars**, and so-so with others. Likewise, some guitars sound great on almost any amp, and others can be quite picky. Ask the salespeople and fellow guitarists for their experiences and always test amps with your own guitar(s).

- Do you play **acoustic** too? Then see what the amp's clean channel sounds like with your acoustic guitar.

121

8

Bass Amps

In the early days of bass amplification, bass amps were little more than guitar amps with bigger speakers, but things have changed. Today, there's a huge range of bass combo amplifiers, heads, and cabinets available, with solid-state or tube technology, or both, with very low and incredibly high power ratings, with a variety of speakers in all sizes, in ported and non-ported designs, with extended EQ controls, and more.

For many years, bass amps didn't do much more than simply make the bass sound louder, primarily producing dark and boomy notes, which was exactly what the music of the time called for. As playing techniques evolved, and as bassists wanted more tone, new generations of amps were developed, providing them with a strong mid and high-end range in addition to a powerful low end.

Any style
Most bass amps do well in any type of music. Still, there are considerable differences from one brand to another: Each company seems to have its own preferred sound, and some companies make amps that are very popular in a particular style of music, such as funk or jazz.

Double basses
Double bass players and acoustic bass guitarists usually go for amps that have a very linear, hi-fi type of character. For specific information on double bass amplification, please see pages 133–134.

Modeling bass (pre)amps
At the time that modeling guitar amps were taking off, modeling bass amps were still scarce: Most bassists tend to go for a single usable sound, rather than for numerous virtual amps and cabs. If you like the amp you have but welcome some variation, you can get yourself a modeling bass preamp (see pages 208–209) on the side.

TUBE OR SOLID STATE

The large majority of bassists use solid-state amps. The 'exact' nature of these amps allows for a faithful, fairly linear reproduction of your input. Also, solid-state amps supply you with lots of power for a reasonable price.

All tubes
Originally, tube bass amps had a notoriously low damping factor, creating a very loose or wooly sound. Today's tube bass amps,

however, can sound as tight as any other amp, and some even have a damping switch that allows you to go back to that old, loose sound. Still, all-tube bass amps are quite rare. Please check pages 100–102 for more information on tube amps.

Hybrid

Hybrid bass amps use a tube preamp only. Besides making for a warmer tone, the tube's high resistance is said to have a positive effect on your overall sound, and to be much better at handling the sudden impulses (slaps, pops) that bass amps have to respond to. And, if you do overdrive the preamp, the resulting distortion will have a pleasant, soft-clip nature (see page 100). The solid-state power amp provides you with a lot of watts at a low weight and a relatively low price.

Distortion

How much distortion you want on your bass tone probably depends on the style of music you play. Some bass amps have a separate overdrive channel for that purpose (see below); other designs use an overdrive or boost button that makes the mid and higher frequencies distort while keeping your low end clean, so you won't lose definition. Speaking of distortion, bassists have made small cuts in speaker cones to produce that effect (the top two speakers of an 8x10 cabinet, for example). This will affect your amp's resale value: Get yourself a bass distortion pedal instead.

Fat

Looking for a really fat sound? The slightest bit of (tube) distortion adds the second harmonic to the notes you play. This frequency, an octave higher than the fundamental, helps create a fatter tone, rather than a really distorted one.

Both

A few bass amps have both solid-state and tube preamps built in, so you can make your choice depending on where or what you play. Some amps even allow you to blend the two.

TIP

> ### Class D
>
> Class D amps have an efficiency of 90% or more. (Class A tube amps hardly ever exceed 25%, while class A/B amps have an average efficiency of some 80%.) This means that they can produce a lot of power without getting really warm, which in turn allows for a very compact design. They're used for some bass amps and PA power amps, for example.

BASS AMP CHANNELS

A very limited number of bass amps — both at the professional and beginner levels — have separate clean and distortion channels, just like most guitar amps. If so, check if you can switch channels with a footswitch. Others have two channels to accommodate two different basses, as many players use both a passive bass and an active bass. (Passive basses don't have a built-in preamp; active basses do, producing a higher output.) Independent tone controls per channel allow you to set the perfect sound for each bass. Check to see if the channels are footswitchable; the switch may be optional.

From one to two
If you have a single channel bass amp and you want to use two different basses without having to adjust EQ and other settings each time you switch instruments, get yourself a footswitchable two channel preamp, each channel providing you with its own EQ and level controls.

BASS AMP INPUTS

Most bass amps have two inputs, one for passive and one for active basses. Their labels are often similar to those on guitar amps (see page 109). Inputs for active basses have a stronger attenuation than inputs for active guitars (-10 or -15dB, rather than -6dB). Again, using a passive instrument in an 'active' input will make for a darker sound, rolling off the high frequencies.

Two basses

Though they haven't been designed for that purpose, you can often use the two inputs to connect two basses as well, or even a bass and a guitar. Don't do this onstage or in a band rehearsal, but it can work out fine at home, for example to figure out a tune with a friend. Of course, the instruments will have to share the tone and volume settings on the amp. If one of the basses is an active model, its tone and level controls may be powerful enough to overcome this.

One input and a switch

Instead of two inputs, bass amps may also have a single input and a passive/active or high-gain/low-gain switch.

Phantom power

A few bass amps use the instrument input to supply *phantom power* (see page 164) for active basses, which need a battery to power their built-in preamp: Just replace your mono instrument cable with a stereo cable, and you never have to change the preamp's batteries again. Of course, your bass needs to be equipped for this.

POWER RATING

Most bassists go for a clean sound at any volume. This, and the fact that boosting low frequencies requires a lot of energy, means that bassists need powerful amps with lots of headroom. For home practice, an amp with a single 6" to 10" speaker and 15 watts or less will do. To keep up with an (unamplified!) drummer and a modest guitar amp, you will need at least 50 to 100 watts. For small gigs, 100 watts may be enough if the music isn't too loud, but anything larger or louder will require 200 watts or more. Bass guitars with a low B-string require special attention: An under-

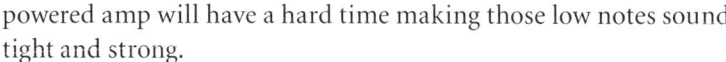

powered amp will have a hard time making those low notes sound tight and strong.

More
If you really want lots of power, it's good to know that there are bass amps with 1,000 watts or more, supplying you with all the headroom you need for an undistorted sound at high volume levels. A few powerful amps feature bi-amping.

Wider sound field, more power
If a bass combo has a jack for an extension cabinet, hooking one up will provide you with both a wider sound field and more power (see pages 84–85)!

BASS SPEAKERS

Most bass combos come with a tweeter and either a single 12" or 15" speaker or two 10" speakers. A 12" is faster and more direct than a 15", but the latter usually has more low end. A few serious combos feature a single 10" or even two 5" (!) speakers for a very tight and well-defined sound. A good speaker of that size can even make a low B sound good.

Large or small
While larger speakers are generally better at producing low frequencies, many bassists prefer smaller speakers. Two 10" speakers move almost as much air as a single 15", but they have better definition, a faster attack, and a tighter sound. Contrary to a 15", they also reproduce your high frequencies, making your bass sound full and complete.

Two, four, eight
Combos and cabs with two small 10" speakers are very popular, but they're usually more expensive than models with a single 12" or 15". There are also cabs and combos with four 10" or 8" speakers, as well as 8x10 or 8x8 cabs.

128

Large and small

Still, large speakers add sub bass frequencies that smaller speakers can't reproduce. As most single big speakers lack the definition you need to hear yourself onstage, they're often combined with one or more smaller ones. Many players use a separate amp head and two cabinets; a 1x15 and a 4x10 is a popular combination. Replacing the 4x10 by a 2x10 makes this kind of setup more portable. There are cabinets that house these and many other combinations in one enclosure too, such as a 12" with four 8" speakers, or a 12" and a single 6".

Tweeter

Most bass combos and cabs have a tweeter built in. Because some bassists find this sound too harsh or brilliant, the tweeter usually has a level control or can be turned completely off. For a milder high end, you may also check out models with a really small (e.g., 5") speaker instead of a tweeter.

TIP

Strong magnets

Good bass speakers have a rigid chassis and suspension, and preferably a long voice coil that allows for long cone excursions. The latter requires a strong magnet that is capable of controlling the cone's movements at all volume levels and the strongest peaks.

PORTED DESIGNS

Many bass combos and cabs have one or two round or rectangular openings in the back or the front of their enclosure. These *ports* don't just let the air move in and out: The idea is to have the back of the loudspeaker make the air in these vent(s) vibrate, producing additional sound energy, an increased bass response, and an extended bass frequency range. A well-designed port also allows the speaker to make smaller excursions, improving power handling and helping to prevent distortion.

129

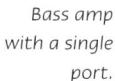

Bass amp
with a single
port.

Front or back

Ports can be front or rear mounted. Front-mounted ports require
more space as they add to the cabinet's diameter. *Tip:* Rear
mounted ports can make for a boomy sound if the combo or cab
is placed too close to a wall — which also means that putting it
closer to the wall can boost low end response.

Standing waves and moving air

Ports with a wedge or conical shape prevent standing waves in the
vent. Larger port dimensions and rounded edges help reduce the
noise created by air that moves in and out.

Pluggable ports and jacks

Some cabinets come with plugs to stop or reduce the venting,
allowing you to 'tune' it to your liking. Undesirable venting
may occur through the jacks. Some amp makers prevent this by
locating the jacks over a sealed chamber, and there are special
plugs that you can use to temporarily seal the jacks.

Tipcode AMPS-012
This Tipcode demonstrates the effects of various EQ settings on your bass sound.

BASS AMP EQUALIZERS

Basic bass amps usually have a three- or four-band rotary EQ, the latter with separate low-mid and high-mid controls. More advanced amps often have extended midrange controls. This is either a graphic or a (semi)parametric EQ.

Midrange
The midrange is essential to the tone of the instrument: The most influential harmonics are in this range. Boosting the high mids makes for an edgier sound, eventually adding an aggressive twang, while boosting the low mids a bit results in a rounder, fuller, fatter sound. For an articulate and tight slap sound, the mids are scooped (and lows and highs may be boosted). The range of adjustable midrange frequency bands varies per amp, of course, from as low as 50Hz (which is actually in the bass range) up to 3 or 4kHz.

Shape
A growing number of bass amps feature a shape or contour control. These controls add color to the tone, offering a variety of pre-shaped timbres: a transparent slap sound, a deep reggae tone, an ultra-low house thump, a heavy rock feel, or a 'fretless' timbre, for example. Many players use their shape control to dial in their favorite tone, using the EQ controls to adapt the sound to the room's acoustics, or vice versa. Also experiment with the shape control at various EQ settings, just to find out where it can take you.

131

Additional EQ controls

When looking for a bass amp, you will also come across an array of additional switches that cut lows (for a cleaner low end); boost lows, mids or highs; add punch or attack; or provide instant scooped mids. Basically, these are single-button 'pre-shape' controls.

Bypass

A (foot)switch to bypass the EQ settings can be used to go from a linear or 'flat' sound to a scooped sound for slapping, or from a linear sound for your electric bass to an EQ-ed sound for your upright, for example.

MISCELLANEOUS FEATURES

The bass is a very dynamic instrument, capable of producing anything from the softest touch harmonics to sudden spikes. That's why many bass amps have a built-in *compressor*: a device that boosts quiet sounds and reduces loud ones, acting as a sort of automatic volume control (but there's more, as you can read on pages 190–192). If you're not a truly experienced bassist, compressors can help you produce a steady sound level.

Dual band

To balance out the volumes of both low and high notes, some advanced bass amps have a *dual band* compressor that processes those ranges independently. The more advanced — and expensive — the amp, the more likely it is that you can control one or more of the compressor's parameters (see page 191).

Limiter

Instead of (or next to) a compressor, bass amps may have a *limiter*, which prevents loud signals from causing distortion or damaging the speaker.

Other effects

Some bass amps have additional built-in effects. For example, there are models that feature an octaver, while others have a number of digital or analog effects on board, such as chorus, flanger, phaser, wah, and synth bass.

Effects loop

Using effects such as a chorus or a flanger usually affects the low frequencies that you need for a well-defined, tight bass sound. To prevent this, a bass amp's effects loop may have a bass cut switch, allowing those low frequencies to bypass the loop unaffected.

Variable crossover

A *variable crossover* is quite a rare feature. It allows you to send more of the lowest bass frequencies to a 15" speaker, and direct all or most of the upper bass and higher frequencies to a 2x10 or 4x10 cabinet, for example, making optimum use of each speaker's size. Keeping the lows away from the 10" speakers gives you more headroom, and feeding just lows to the 15" will make it sound cleaner.

DOUBLE BASS AMPS

There are a number of 'regular' bass amps that are favored by double bassists too, and a good salesperson can tell you all about the current models. Other amps have been specifically designed for the double bass (*a.k.a.* acoustic or upright bass). They're used by acoustic bass guitarists as well.

Small

Double bassists often go for relatively small solid-state amps rated at 100 to 200 watts. The sound should be as natural and 'acoustic' as possible. To emulate the omnidirectional nature of the double bass, some models have a *downfiring* woofer that uses the floor to reflect the sound in all (*omni*) directions. Rather than a tweeter, double bass amps often have a second, smaller speaker for the mid and high ranges.

Preamp

A (piezo) pickup with an ultrahigh impedance is a popular choice

133

A 340+60 watt
(double) bass
amp in
a 16x16x14"
enclosure
(AER).

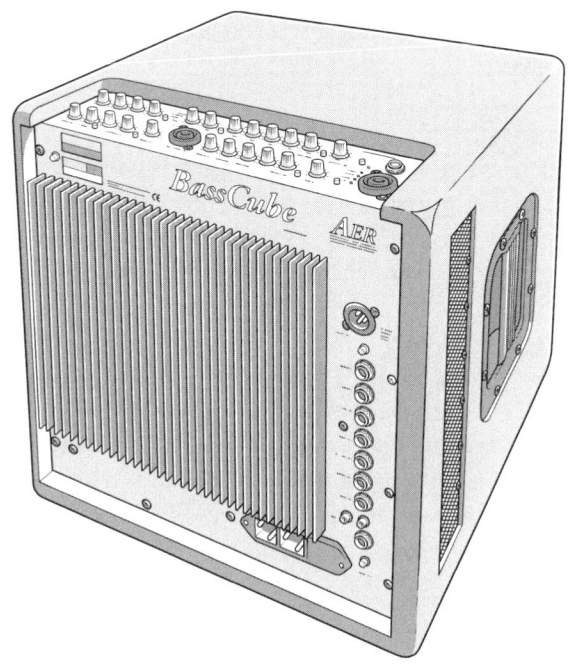

among double bass players. Connecting such a pickup to a regular high-impedance instrument input, as commonly found on bass amps, will degrade the signal. This can be solved by using a special type of preamp that lowers the pickup's impedance level. These preamps are small boxes, usually with a belt clip, so you wear them on you. This has the added benefit of reducing the distance between the pickup and the preamp: ultrahigh-impedance connections are sensitive to hum, and a short signal path helps preventing unwanted noise.

Acoustic amps
Alternatively, there are bass amps and acoustic amps that have an ultrahigh-impedance input. The next chapter tells you all about acoustic amps.

Microphones
Rather than a pickup, some upright bassists use a mic, or a pickup/ mic combination. To accommodate them, some bass amps have a dedicated microphone input.

134

> ### Feedback
> As double basses tend to feed back, a notch filter and/or a
> phase switch is a welcome feature on a double bass amp or
> preamp. See pages 142–143 for more information.

TIP

Powered speakers
If you're going to get a separate preamp, you may consider one that
has volume, tone, and other controls. With those features, you
can combine it with one or two powered speakers, rather than a
bass amp. Powered speakers (see pages 164–168) are designed to
produce a very linear, uncolored sound, which is what most double
bassists are after.

TESTING BASS AMPS: TIPS

- Does the amp produce a powerful, tight, and balanced sound at
  low volume levels, should you intend to use it that way?

- Does the volume level continue to increase when opening up
  the **volume control** to the level you require?

- Does the amp still produce **a clean sound** at that level? Also
  check that nothing other than the store's inventory rattles at
  high volume settings.

- A bass amp's neutral **EQ settings** best reflect the sound its
  designers had in mind, so start there. If you have an active bass,
  set its EQ controls in their center (flat) positions too. Passive
  bass? Turn your tone controls wide open.

- Low E has a wavelength of more than 27 feet (8.33 meter), and
  a low B wave is over 36 feet long. To fully appreciate the sound
  of a bass amp, **move away from it**: Have someone else play, and
  judge the amp's performance from a distance.

- If your low E and B lack punch and definition on most amps,

135

you may want to try **a different bass**, rather than blaming the amps.

- If you play **a variety of basses**, try them all out. Some amps stress the differences between fretted and fretless basses, others seem to disguise them.

- Note that amps with a very articulate nature may be quite **unforgiving**: They make both intended and unintended subtleties come out.

- Use your **entire dynamic range** (from touch harmonics to slaps and pops) and see how the amp reproduces your playing. Some amps seem to need a bit of tweaking when you switch from slapping to fingerstyle playing, while others cope with both techniques equally well without any adjustments.

- Are you a **multi-instrumentalist**? See what your bass amp sounds like with a keyboard or your acoustic guitar.

- Again, listen to the **overall sound** as well as to the character of the lows (fundamental, fat and full, tight and clean, and so on), the mids (punchy, boxy, pronounced, strong, weak, etc.) and the treble (bright or harsh, edgy or mild, detailed, hazy, etc.).

9

Acoustic Amps

Though most acoustic amps are primarily designed for acoustic/electric guitars, they're used for a host of other instruments as well, including vocals, electric violins, and electric guitars.

The highest compliment you can pay to an acoustic amp is that you don't notice it's there: The instrument sounds exactly the same as if it were unamplified, although a little or a lot louder. Most bass and keyboard amps are supposed to do the same. So what makes acoustic amps different? They often have special inputs for ultrahigh-impedance pickups, as well as one or more microphone inputs. They also usually have smaller speakers, their built-in effects are geared toward acoustic instruments, and they have special features to fight feedback — a common problem with the instruments that these amps are used for, from guitars to mandolins and banjos.

The instruments
Acoustic amps can also be used to amplify horns, vocals, and lots of other instruments, including upright bass and acoustic bass guitars. Acoustic guitarists who double on electric (but don't want to lug two amps around) have had success using an acoustic amp with a

An acoustic amplifier. (SWR)

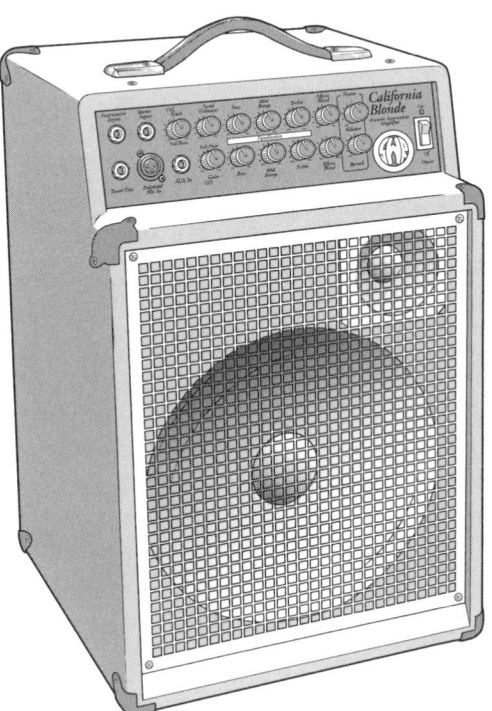

distortion pedal, a fuzz pedal or a modeling preamp. Again, some amps lend themselves better to these experiments than others. Jazz guitarists and other 'electric' players who're looking for a clean tone often go for acoustic amps too. (Some plug their guitar straight into the mixing board for an even cleaner sound.)

Sound system

You can also use two acoustic amps as a stereo sound system for small venues, just like you can with keyboard combo amps. Do note that the equalizer of an acoustic amp is typically designed to shape the sound of acoustic instruments, or even just guitars, rather than the sound of a singing voice.

POWER RATING

Power ratings for acoustic amps range from a mere 10 watts for practice amps to some 200 watts. To play larger venues, the amp's signal can be fed to the PA (provided it has a DI or a line out), while you use the built-in speakers for personal monitoring. You can also plug directly into the mixing board, usually through a DI box, especially if the board is far away.

How much

Generally speaking, a good, efficient 60 or 100 watt amp is enough to cope with a big band, fill a small club, or play with some 500 people singing in a church. If you want to use it for an upright bass as well, you'll need to go for more power (and usually for larger speakers too). Most popular acoustic combos seem to be in the 40 to 80 watt range.

Small and lightweight

Users of acoustic amps commonly prefer small, lightweight amps, and you can get 60 watts of clean power in a box that measures no more than a cubic foot and weighs less than twenty pounds (9 kilos). The majority of acoustic amps use lightweight solid-state technology.

139

SPEAKERS

Most acoustic amps have smaller speakers than other types of amps, from one or more 5" speakers to 8" dual-cone designs. If you want to use the amp for an electric (bass) guitar or a double bass as well, you'd probably better look for models with a 10" or a 12" speaker.

Tweeter

Most acoustic amps have a tweeter, or even two for an improved dispersion of the highest frequencies. As the tweeter can make some pickups (particularly piezos) sound quacky and violins sound harsh, it can usually be disabled or attenuated. If so, you can also use this *high-end pad* to reduce excess fret, string, bow, or finger noise.

TIP

> ### Extra speaker, more power
> If an acoustic combo has a jack for an extension cabinet, hooking one up will provide you with more power (see pages 84–85)!

Open or closed

Just like electric guitar amps, acoustic amps come with open or closed backs. The difference between the two is explained on pages 114–115.

CHANNELS AND INPUTS

A two-channel amp can be used to amplify two instruments, or your guitar and your vocal mic, or an instrument with a combined pickup/mic system (see page 227). Independent level and tone controls allow you to balance and tweak the signals from both sources. Want more? There are four-channel acoustic amps too.

Features

The usability of the channels depends on their features. Some amps provide separate instrument and microphone/auxiliary channels, while others have two or more identical channels, both with microphone and instrument inputs. If you plan to use the amp for two acoustic instruments, it may be good to have separate notch filters per channel. Advanced models have independent effects loops too, so you can add different effects to your sound sources. A separate electric guitar input may be provided as well.

Inputs

The wide variety of instruments that acoustic amps can be used for requires inputs that accept a wide variety of signals, from low-impedance mics to ultrahigh-impedance pickups. This can be done by providing different inputs or switchable inputs, and some amps have *buffered inputs* that can handle various signal and impedance levels.

Preamp

Connecting an ultrahigh-impedance pickup to a regular instrument input degrades the signal, as explained on page 88. This can be solved by using a special preamp. Most newer acoustic/electric guitars that feature such a pickup have one built in. Separate preamps are available in various formats. Some musicians prefer a small, battery-powered belt-clip model that allows you to control things from your playing position.

Preamp features

Preamp features can include tone controls, gain controls, notch filters, and phase switches. For combined pickup/mic systems, you also need a second channel, as well as a control to set the balance between the two sources. Some preamps double as DI boxes, allowing you to connect your instrument straight to the mixing board.

Microphone inputs

Microphone inputs are usually designed for professional mics (lo-Z, XLR connector). Check to see if they supply phantom power, allowing you to use condenser mics (see page 217).

141

'ACOUSTIC' EQUALIZERS

Most acoustic amps have a basic three-band rotary EQ, which is typically used to adapt the tone to the sound of the room you're playing. Some have additional controls to color or shape the sound.

Color

These controls usually offer one or more EQ 'presets.' For example, there are shape controls that boost the low and high end while rolling off the midrange, producing a timbre that is designed for strumming. Other switches just add a bit of brightness to the sound, or they boost the mids for fingerstyle playing. More advanced shape or enhance controls offer a number of variable tone settings or curves, typically affecting different frequency ranges than the EQ's tone controls do.

FEEDBACK

A notch filter is actually a sweepable tone control, meant to filter out just the feedback frequency. This implies a very narrow bandwidth: The narrower the affected frequency range, the less the filter will change the overall sound. If a notch filter has a pronounced effect on the sound, the cure may be worse than the disease: Notch filters can cut up to 20dB.

Automatic

Rather than searching for the offending frequency yourself (turn the notch filter off, then crank up the volume until feedback occurs, and dial it out using the notch control) some amps have a circuit that automatically locates it.

Phase switch

If the amp has a phase switch or *phase inversion switch*, first see what this does to reduce feedback. After all, using the phase switch

142

will not affect the overall sound at all. Note that the same switch can be used to get signals from an instrument with both a pickup and a microphone back in phase, improving the sound. If signals are out of phase, they will partially cancel each other out.

Still feeding back?

Some amps seem to be more sensitive to feedback than others, especially when used with certain types of guitars — so it's best to bring your own instrument when you're going to audition amplifiers. If feedback problems persist, try some of the suggestions on page 247.

TIP

MISCELLANEOUS

Reverb is a very common effect in acoustic amps. Unlike the spring reverb in electric guitar amps, an acoustic amp's reverb is usually digital. Chorus is another popular effect, though players who go for a true, natural sound usually don't like it. Certain acoustic amps feature built-in delays too, or even a complete multi-effects unit. On some amps, the effects can be turned on and off with a footswitch.

Headphones?
A headphone output on an acoustic amp might seem kind of weird: If it's too loud, you just turn the amp off, don't you? Still, it definitely makes sense for electric violins and cellos or for solid-body acoustic guitars; or to use effects at low volume levels; or for monitoring your music while the amp is connected to a recording console.

Modeling
Guitars (and other acoustic instruments) often sound best with a microphone rather than a pickup, but microphones are usually very impractical for onstage use. They limit your freedom of

143

movement and easily cause feedback. A microphone modeling unit may help: The modeling technology shapes your pickup's output signal as if you were using a genuine mic. Though usually designed for acoustic guitars (they can even make a steel-string sound like a nylon-string guitar), you can use them for other instruments as well.

TESTING ACOUSTIC AMPS: TIPS

- Most players want acoustic amps to sound as natural as possible. You can check this by playing while someone else turns up the volume from zero, **very, very slowly**. With a good amp, it'll be hard to tell when it gets louder than the instrument.

- Acoustic combos are supposed to sound **transparent, open, and musical**, and to respond to every nuance in your playing. They should be crisp but not overly crisp, bright but not harsh or tinny, warm but not wooly, and clean without being sterile. Listen to every element in the sound at all volume levels.

- Don't overlook the smallest amps: Some can sound surprisingly **big**.

- How sensitive is the amp to **feedback**? Turn the volume up and see what happens, and check how well you can fight feedback with the available controls.

- Acoustic combos are sometimes said to make **lesser quality instruments** sound better, adding personality, depth, or richness. If an amp does so, it may overdo things with a great instrument, resulting in an exaggerated low end, for instance. Again: Audition amps with your own instrument and the pickup or the mic you normally use, or with the one you plan to buy soon.

10

Keyboard Amps

At more than seven octaves, the piano and the organ have the largest range of all acoustic instruments. Digital keyboard instruments can produce even lower and higher frequencies, as well as any sampled sound you can imagine, from strings to drum machines. No wonder keyboard amps are often referred to as 'full-range amps.'

Keyboard amplifiers are true multi-purpose amps. That said, they can't provide you with the tone and the distortion of a guitar amp, or the power and punch of a dedicated bass amp, but they can definitely reproduce those sounds, just like a home stereo system or a PA can.

Sound system?

You can use keyboard amps as a miniature sound system, albeit with limitations: Most models don't have independent tone controls per channel, or many of the other features that dedicated sound systems provide. Also, they usually have just a single mic input.

A very basic keyboard amp.

Alternatives

At live performances, keyboard players often use the sound system. You can either connect the instrument to the mixer or use the line outputs or DI of your keyboard amplifier, so the latter can double as a personal monitor. There's more on sound systems in the next chapter. (*Tip:* The powered speakers mentioned in that chapter can also be used to amplify your keyboard instrument!) At home, you can use the keyboard's built-in amp and speakers (if

146

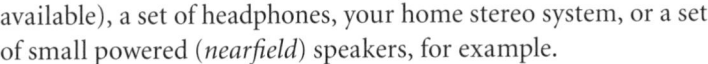

available), a set of headphones, your home stereo system, or a set of small powered (*nearfield*) speakers, for example.

Small range

The range of available keyboard amps is smaller than the range of, say, guitar or bass amps. Yet, there are plenty of models to choose from if you want a truly portable and multi-functional system that you can use for both practice and performances.

SPEAKERS AND POWER RATING

The smallest keyboard amps often have a 10" speaker and a tweeter, or a dual-cone speaker that performs both functions, and a power rating of 15 to 30 watts. For small venues, most players go for 60 to 100 watts and a 12" speaker, or two smaller speakers, as well as a tweeter. Top-of-the-line keyboard amps usually have a whole array of speakers, making sure they fully cover every frequency range: a 15" woofer, one or more 6" to 10" midrange speakers, and one or two tweeters. The most powerful models can put out up to 1,000 watts.

TIP

> ### Subwoofer
> *Do you want more bass than your keyboard amp delivers? If it has a line output, you can hook up a powered subwoofer. Dedicated keyboard subwoofers usually have a 15" speaker and around 200 watts of power.*

Two times mono

As said before, most keyboard amps are mono. For a stereo effect, you can often connect a second amp in a way that the overall volume can be controlled from the first amp.

Stereo

Some companies make stereo keyboard combos. To create a stereo image, they have an angled cabinet design, one speaker facing a bit to the left, the other to the right. To enhance the stereo image, they may have a *spatial expander* or a similarly named effect.

CHANNELS

Each channel usually has a separate level control and possibly some other knobs, such as EQ or effect level controls, but the latter are usually shared. Of course, you can also connect two mono instruments to one stereo channel, but they will have to share all that channel's controls.

How many
The number of channels usually goes up with the amp's power rating and price. Four stereo channels will be sufficient for most applications.

Microphone
One channel usually has a balanced microphone input, preferably with an XLR socket. Keyboard amps with two or more microphone inputs are rare.

Tube preamp
Tone wheel organs and various electric piano sounds can be enhanced by the warmth — and, if so desired, the characteristic distortion — of a tube preamplifier. Rather than buying one separately, check out the rare keyboard amps that have a special channel with a tube-driven preamp.

TIP

Electric pianos and guitar amps
For their vintage electric pianos some players prefer using guitar amps rather than keyboard amps, for example connecting a Fender Rhodes piano to Fender Twin or Roland Jazz Chorus amp.

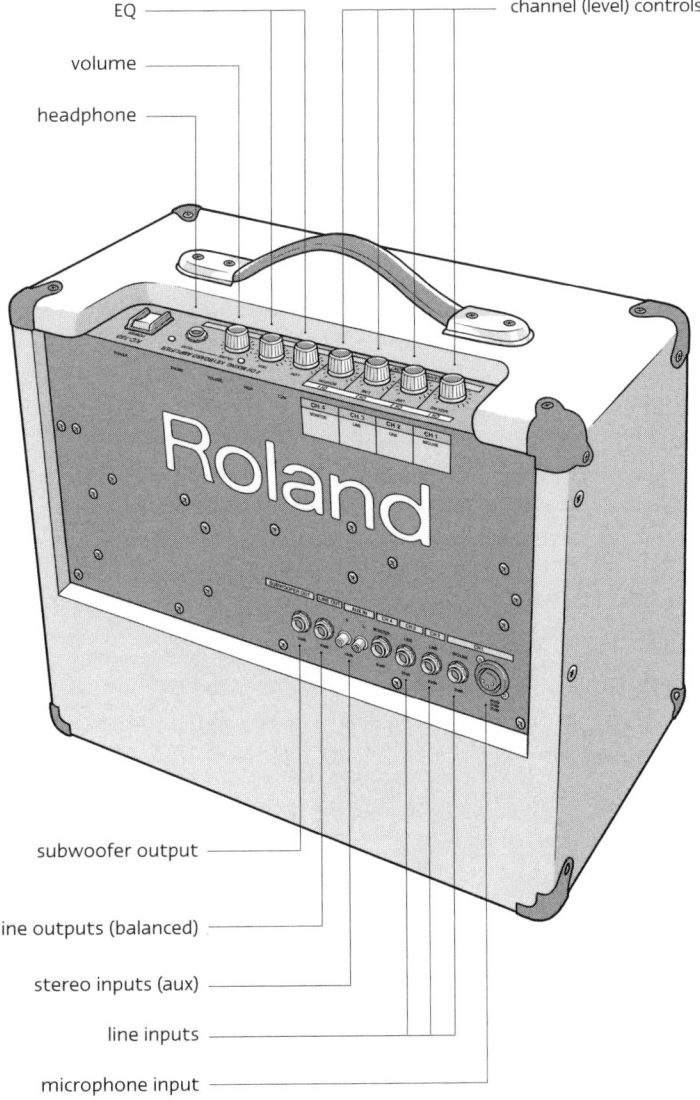

EQ

volume

headphone

channel (level) controls

A keyboard
amp.

subwoofer output

line outputs (balanced)

stereo inputs (aux)

line inputs

microphone input

Organ channels and cabinets

Other keyboard amps have a special organ channel that emulates the sound of classic organ speaker cabinets, such as the famous Leslie cabinet. This classic powered enclosure features a rotating horn for the treble range and a rotating drum that swirls and projects the woofer's low frequencies. Leslie is a registered

149

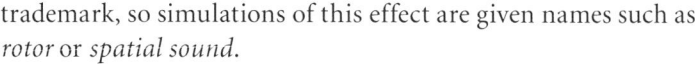

trademark, so simulations of this effect are given names such as *rotor* or *spatial sound*.

MISCELANNEOUS FEATURES

You may also come across the following features.

- Most keyboard amps have **few or no effects** built in, mainly because keyboard instruments often do. An effects loop may be present, as well as a reverb, helping to bring vocals and other instruments to life. Some combo amps have a whole range of digital effects built in. The effects may be footswitchable, and they may be assignable and/or adjustable per channel.

- Keyboard amps often have an **auxiliary input** for CD players or similar sound sources; they're perfect for playing back pre-recorded music between sets, or in a DJ setting.

- Keyboard amps preferably have two balanced DI outputs or a stereo line out, so you can send a stereo signal to the PA or recording console.

- For sound system use, check to see if the amps have speaker stand inserts (see page 39).

Testing tips

For keyboard amps, all the general testing tips on pages 91–96 apply. One extra tip: When testing keyboard amps, try them with all the instruments and sounds you plan to use them for, including vocals.

11

Sound Systems

Sound systems really come in all sizes, from compact,
portable packages to systems that fill stadiums with
hundreds of thousands of watts. This chapter focuses
on smaller systems that are bought by musicians rather
than by sound engineers. These systems are also used
to amplify speech and pre-recorded music in conference
rooms, churches, schools, and other venues.

The mixer is the heart of the sound system. The power amp, another essential component, can be separate, or built into the mixer or the speakers. As with all other systems, separate components offer the most flexibility: You can choose your own combination of mixer, amp and speakers. If you need bigger speakers, a more powerful amp, or a mixer with more channels, you can buy just that.

Integrated
On the other hand, systems with an integrated amp are easier to set up and break down, there's less hassle with cables, and therefore less chance of noise, hum, and other problems.

Powered speakers or powered mixer?
Powered mixers have been around for years, whereas powered speakers became increasingly popular and affordable around the turn of the century. Price-wise, the difference between these two options has gotten smaller and smaller.

The difference
Powered speakers or *active speakers* have a number of advantages over powered mixers:

• The built-in power amp is **designed for the speakers**.

• The amp is as **close to the speakers** as possible: the signal path is very short — and the shorter, the better.

• You can **daisy-chain** any number of powered speakers.

Heavier
That said, the built-in amp makes powered speakers heavier and therefore harder to handle than 'amp-less' *passive speakers*.

More power
If you really need a lot of power, you may need a separate power amp. Powered speakers typically don't go beyond 400 or 500 watts (see pages 166–167). There are powered mixers with higher power ratings (from 2x75 to more than 2x750 watts), but the most powerful models may very well have more channels than

152

you'll ever need, which is not really efficient. Separate power amps are available with as many watts as you need, and they can be combined with any size mixer.

Lower is more

The number of watts produced by a power amp changes with the impedance of the speakers you use. Power amps can often handle speaker impedances as low as 2Ω, implying that a relatively 'light' amp can produce an awful lot of power. For example, a power amp that supplies 2x400 watts at 8Ω, may generate 2x1,000 watts at 2Ω. Used in the bridged (mono) mode, you're talking about 2,000 watts of power.

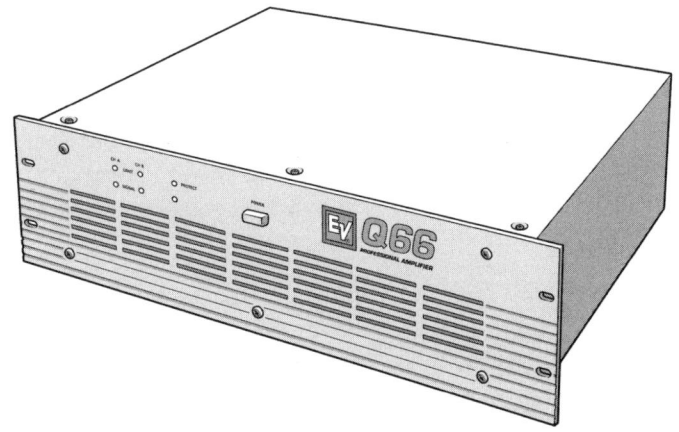

A power amplifier.

How much?

So how much power do you need? The size of your audience is one element. There's a rule of thumb that says that 10 watts per person will do, but this strongly depends on the music you're playing: An audience of five hundred that sits down to listen to a jazz quartet requires a less powerful system than a head banging crowd of a hundred. Ask an expert for advice, and you will be asked what style of music you play, what the largest venue is that you will play with the system, which instruments you plan to use the system for, and so on.

Power

Two speakers and 100 to 150 watts per channel may do if the

153

system is used for vocals only, assuming the drums are not miked and the rest of the band uses their own amps (the *backline*). If the drums need to be miked and the bassist is playing through the system too, you will soon need a system that you'd rather rent than buy — and have a sound engineer come along, rather than handling the controls yourself. The same goes if you play larger places. A convincing sound level in a 2,000 square foot room might require 10,000 watts of power.

THE MIXER

Powered and non-powered mixers are available in three basic designs:

• **Box type mixers** have their controls on a recessed front panel, making them less vulnerable to spilled drinks and other damage. Closing the box after the show protects the mixer on the road.

• **Tabletop models** have their controls on top. Large sound

*A powered
box-type mixer
(Phonic).*

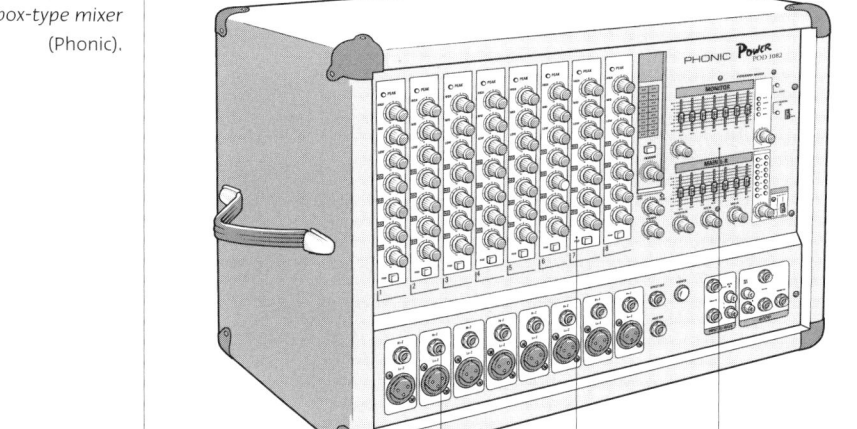

input
section   channels   master section

systems and studios always use tabletop models. Sockets can be rear-mounted or top-mounted. Top-mounted sockets may look messier, but they let you plug things in and out from your working position.

- **Rack-mountable mixers** are available too. Some tabletop mixers come with *rack wings* to make them rack mountable.

Two sections

Due to their large number of controls, mixers can look pretty intimidating. They're not that complicated though, once you divide them up in their two main sections: the channels, and the master section.

Channels

The channels take up the largest part of the mixer's surface. Take a closer look and you will see that each channel consists of a number of identical controls, with one (mono) or two (stereo) inputs at one end, and a fader at the other, as shown on page 156. Every instrument and every microphone is connected to one channel, the stereo channels being used for stereo instruments such as synths and digital pianos. This allow the engineer to adjust the sound of each instrument and voice, using the faders to create a balanced mix.

Master section

This mix is fed into the mixer's master section, where the overall volume is set, among other things. The controls of the master section affect the combined signals of all channels.

How many?

The number of channels ranges from two to twenty or more. How many channels you need is hard to say, but it's always good to have a few channels more than you think you need today.

Channel controls

On any given mixer, each channel basically has the same controls and connectors. Here's what you will commonly see, from top to bottom:

155

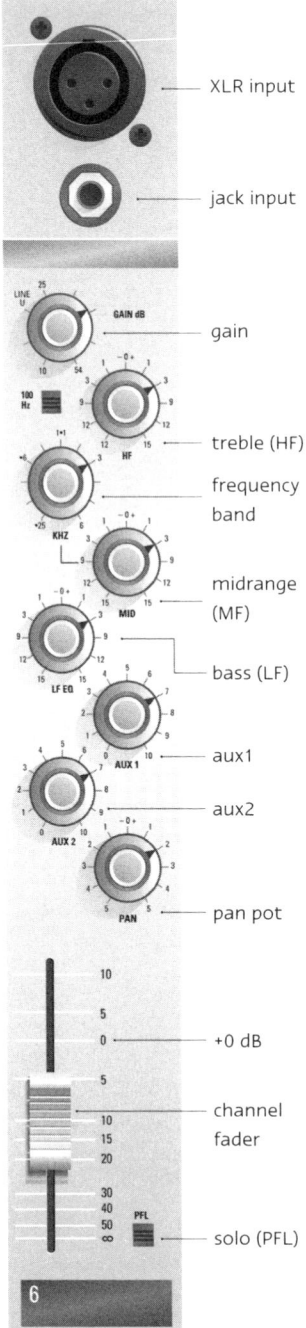

A channel strip and its inputs and controls.

XLR input

jack input

gain

treble (HF)

frequency band

midrange (MF)

bass (LF)

aux1

aux2

pan pot

+0 dB

channel fader

solo (PFL)

- Channels usually have both an **unbalanced jack and a balanced XLR** socket, or a combined socket for both types of plugs.

- The **channel gain** or **trim control** sets the input level for each instrument or microphone.

- Each channel has its own **tone controls**.

- One or more **aux controls** allow you to adjust effect levels per channel, or to adjust the volume level of an instrument in the monitor mix (see page 161). If a mixer has built-in effects, there may be dedicated controls to set the effect level per channel.

- A panoramic potentiometer or **pan pot** allows you to move the channel's sound from left to right in the stereo image.

- The **channel fader**, a sliding control, is used to adjust the volume level of each instrument or microphone. Small mixers sometimes have a space-saving rotary level control instead of a fader, saving both room and money.

Stereo channels

Most mixers have one or more stereo channels for keyboards and other stereo instruments, stereo effects units or a CD player. To allow for the different signal levels

156

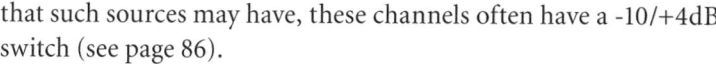

that such sources may have, these channels often have a -10/+4dB switch (see page 86).

One stereo, two mono

The inputs of a stereo channel can also be used as two mono inputs, but then you won't have independent control over those two sources. If a 12-channel mixer has two stereo channels, they're usually numbered 9/10 and 11/12. *Tip:* Stereo channels don't have microphone inputs.

Gain control

Setting the gain control is no different than on a combo amplifier (see pages 58–59). There may be a peak indicator per channel, or you have to watch the mixer's *LED ladders* (see page 159).

Channel fader

Once the channel gain controls have been adjusted, the channel faders will usually be set around 0dB, which is usually about two-thirds of the way up. This 0dB setting represents the channel's *nominal level.* This is where you have the optimum signal level: high enough to prevent hiss and low enough to prevent distortion. This position also leaves you plenty of *fader throw* to adjust the channel's volume level in the mix.

Fader throw

The larger a fader's total throw, the more finely you can adjust it. Fader throws vary from less than 50 to 100mm (about 2" to 4"), the latter being referred to as a *full-range fader.*

Panorama or balance

The *panorama* or *pan pots* allow you to separate instruments in the mix, rather than making them sound as if they all come from the same spot. Live, the pan controls are typically used to duplicate the stage image: If the horn section is on the far left, their sound should come from the far left too. On stereo channels, the pan control is replaced by a *balance control.*

Tone controls and filters

On many mixers, the frequency band for the treble control is

157

around 10kHz, but other mixers use 12k for the treble. The bass control is usually at 100Hz or a bit lower, at 80Hz. The midrange can be anywhere from 1 to 2.5kHz, and there are mixers that provide you with a sweepable midrange EQ. To fight stage rumble and other unwanted low frequencies, mixers may have a *low-cut* (or *hi-pass*) filter per channel.

Solo

If you want to check the EQ or any other settings of one of the channels without hearing the others, you can press a small button that's commonly labeled *solo*. It is usually located near the channel fader. This button is also used to check the channel's gain setting. For that purpose it should provide a signal that's independent of the position of the fader. This is known as *pre-fader listening* or *PFL*.

Check this channel's EQ, gain, and other settings with the solo (or PFL) button. The mute button allows you to mute the entire channel.

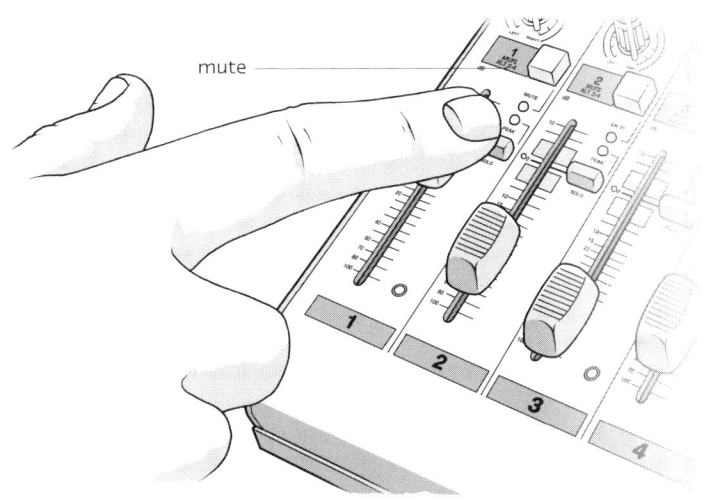

mute

After-fader

If you want to check the volume level and the stereo position of a channel in the overall mix, or if you want to hear how *after-fader effects* respond on the fader's settings, you will also need *after-fader listening* or *AFL buttons*. Most mixers have either just PFL or (switchable) PFL and AFL buttons.

MASTER SECTION

The master section controls the overall output. There are one or two master faders that control the amount of signal that is sent to the power amp. Of course, a single fader doesn't allow you to adjust the left-right balance. Other faders or controls in the master section may control things such as the volume level of the monitor mix, the master level of built-in or external effects, and/or the level of a dedicated subwoofer output.

LED ladders

On most mixers, two LED ladders or *bargraph meters* visually represent the overall signal level, but they may be used to monitor individual channels too, for example when adjusting gain settings. For optimum results, the master faders should be set so that the LEDs light up to the 0dB mark (the nominal level) and slightly above. The ladders usually have three sections, marked with green, orange and red LEDs respectively. To prevent distortion or damage, the red LEDs should either not light up at all, or at very loud peaks only. Check your manual for details.

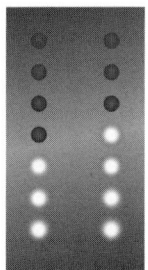

Green orange, and red.

Master EQ

If there's a master tone control, it's usually a graphic equalizer that is mainly used to adapt the overall sound to the room, dialing out boomy lows, overly bright reflections, and other unwanted frequencies, or boosting others for a bit of extra warmth, projection, or sparkle. When there are two graphic EQs, you may be able to use one for the right and the other for the left channel, or one for both channels and the other for the monitor mix.

Reverb

The most common built-in effect is reverb, as this is a typical *master effect* that most instruments benefit from. It will usually be a digital reverb, providing various types of the effect (see page 174) rather than the spring reverb that's often used in guitar amps.

Multi-effects

A growing number of mixers have an integrated multi-effects processor, or even two, the latter allowing you to use different effects simultaneously and to assign different effects to different channels. Common effects are chorus, phaser, flanger, and delay. A compressor is a helpful effect, especially for singers (see pages 190–192).

TIP

> ### Footswitch
>
> **Mixers with built-in effects often have a footswitch that can be used to switch the effects on and off, or to set the delay time by tapping the switch in the tempo of the song (see page 177).**

MORE ON MIXERS

The larger the mixer, the more controls, outputs, inputs, and other features you may come across. Here are some of the main ones:

Auxiliary

Aux or *auxiliary (send) outputs* can be used for effects or powered (monitor) speakers, among other things. When used for an effects device, the effect's output will be connected to an *aux return* input, if available, or to one or two of the mixer's channels. In the latter case, you can EQ the effect's signal before it reaches the master section.

Aux controls

Most mixers have more than one auxiliary output, commonly

labeled aux1, aux2, etc. For each aux output, each channel has an *aux control* that is labeled accordingly. When aux1 is used for monitoring, the aux1 knobs on each channel are used to set the levels of the channels in the monitor mix.

Pre-fader

If you use an aux send for monitoring, you don't want the volume level in the monitor mix (what the band hears) to change when you adjust an instrument's level in the *house mix* (what the audience hears) by moving the channel fader up or down. This requires the aux send level controls to be *pre-fade*. They should be independent of the fader's position.

Post-fader

However, if you use an aux send for an effect, you want the effect to fade up and down with the fader. This means the aux send controls for effects need to be *post-fader*: They should change depending on the fader's position.

Both

So, aux send controls can be pre-fader or post-fader, or they can be switchable, either overall or per channel. The last option offers the most flexibility.

Monitoring

Larger mixers have separate *foldback* or *monitor sends* and a control to adjust the overall monitor volume. If there are multiple monitor sends, you can make various 'personalized' monitor mixes: The drummer usually wants to hear the bass very clearly, for example, but may not care for the horns or background vocals in his or her monitor mix. On the other hand, the vocalist might want to hear the guitar or the piano and the background vocals, rather than the bass.

Monitor power

Most powered mixers have line level outputs for the monitors only. This signal needs to be amplified, so you need powered monitor speakers or a separate amp with passive speakers. Others have an extra power amp built in to drive the monitors. *Tip:* If a system

161

has a 2x200 watt main amp and a single 200 watt monitor amp, it may be advertised as a 600 watt sound system, which makes it seem quite a bit more powerful than it really is.

Your choice

With other systems you may be able to either address the audience in stereo, or to use one of the stereo channels for the front-of-house (FOH) speakers and the other for monitoring, both in mono.

A sound system with monitor speakers.

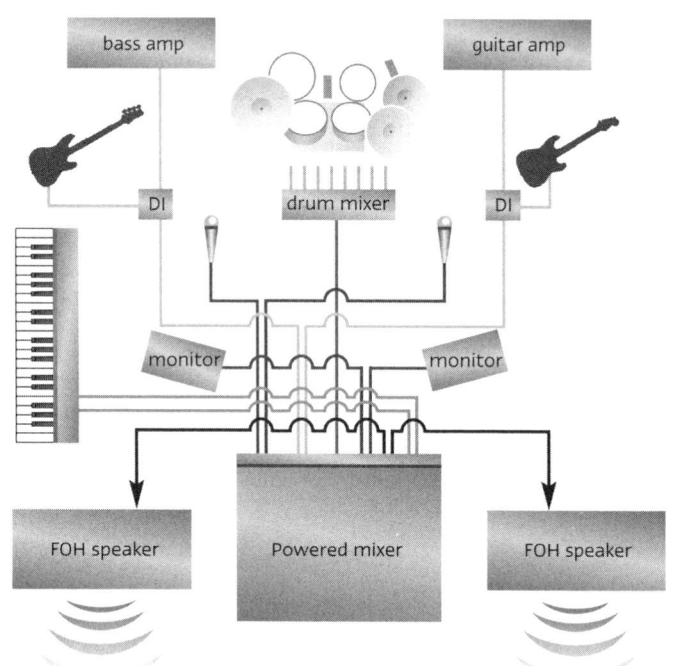

TIP

Recording

You can also hook up a recording device (tape, MD, CD-RW, etc.) to an aux output, but it's easier if the mixer has separate recording outputs. These are usually RCA sockets and/or mini stereo jacks. A tape input with similar sockets may be provided for playback.

162

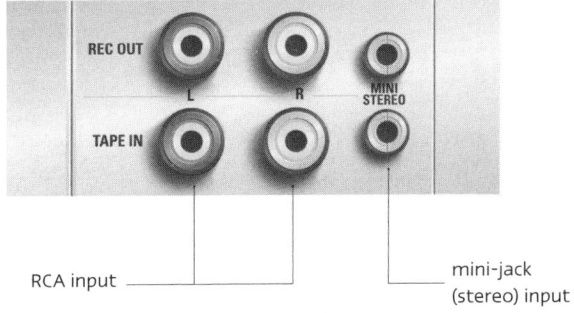

Recording out and tape in, supporting RCA and mini jacks.

RCA input

mini-jack (stereo) input

Insert

An *insert* is a combined input and output that you can use to connect an effects unit such as a compressor or a limiter. Most smaller mixers have one or two master inserts; others have separate inserts per channel. An insert is very much like a serial loop.

Inserts and Y-cables

Inserts require a Y-cable with a stereo (TRS) plug on one side, and two mono plugs on the other. The signal leaves the mixer through the TRS plug's tip, sending it to the effect. It returns to the mixer through the TRS plug's ring.

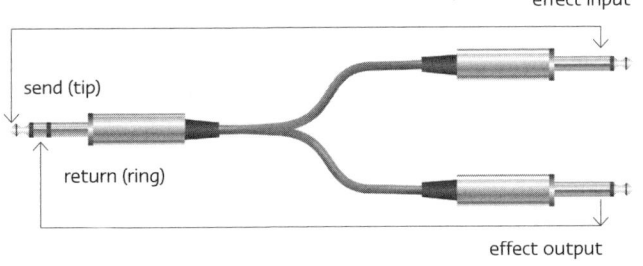

effect input

send (tip)

return (ring)

effect output

A TRS-plug breaks up the signal path at the insert. The signal is sent to an effects unit and then returned to the mixer.

Tip or ring?

In the illustration above, the tip is Send while the ring is Return. This is the American standard. According to the British standard, the tip is Return; the ring is Send.

TIP

163

PHANTOM POWER

Condenser microphones need power to operate (see page 217). This is usually supplied by the mixer. The XLR inputs send power to the mics through the mic cables, so there's no separate, visible power cable — hence the name *phantom power*. (It's just like computers that use USB cables to power scanners and other peripheral devices.)

Per channel or master

Phantom power can be switched on either per channel or globally. The difference? Switching it on per channel reduces the risk of inadvertently sending DC power to devices that may get damaged by it, such as a dynamic microphone that uses an unbalanced (i.e., non-standard) cable. When using standard cables, the risk of things going wrong is minimal.

More or less

Most mixers provide 48-volt phantom power. With lower voltages, some condenser microphones may not work, or they will have limited headroom.

DI boxes too

Phantom power can also be used to power active DI boxes or bass guitars (see page 60), for example.

(POWERED) SPEAKERS

First of all: When you're about to buy a sound system, make sure to spend the largest part of your budget on speakers! Sound systems need full-range speakers that project the sound into the audience, so everyone gets to hear the band. Most studio-type speakers are so-called nearfield monitors, designed for listening at small distances.

PA or monitor?

PA and monitor cabinets are typically two- or three-way systems with a ported design (see pages 129–130). Many designs feature a multi-angle enclosure that lets you use them both as front-of-house PA speakers (the mains) and as wedge-style floor monitors, the latter preferably at dual or triple angles. To make speaker cabinets effective as front-firing mains, they should be pole-mountable: If you want the people in the back to hear your music too, the mains' bottoms should be above the tops of your audience's heads.

Tip: Some cabinets feature a rotatable horn, so you can have maximum horizontal dispersion in any position.

Small

Even cabinets with 10" woofers can be used as front-of-house PA speakers (and there are both entry level and professional quality cabs in that size), but you will need a whole lot of them if you want the lowest frequencies to reach the back of the venue. Some cabinets feature even smaller 8" or 6" speakers: These are satellites that come with a separate subwoofer.

Larger

The 12" speaker is a popular size in PA and monitor speaker cabinets, either with or without a separate midrange driver. Midrange drivers are usually 6" to 8". Subwoofers are often 15" or even 18". That said, there are systems with two 12"s that claim to produce the same amount of lows yet with higher speed, more control, and a faster attack. If you want subwoofer lows in a single-box design, also check out some of the dual 15" speaker systems that use one speaker for lows and mids, and the other as a built-in subwoofer.

Pole-mounting and subwoofers

Adding a subwoofer should also be considered when you're pole-mounting your PA speakers: You need the high and midrange frequencies to project towards the back of the venue, but low frequencies like to travel low. In some systems, the subwoofer doubles as a speaker-stand base. An example is shown on the next page.

The subwoofer doubles as a speaker stand base.
(JBL)

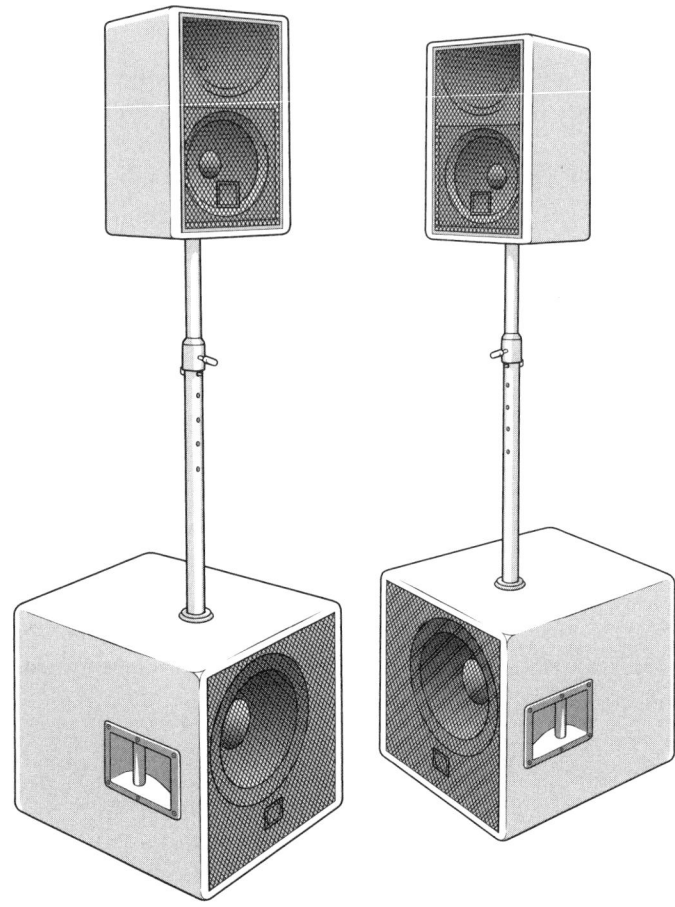

Features

- When buying speakers, check what type of **sockets** they have. Speakon (see page 55) is best, followed by XLR and ¼" inputs.

- Speakers may have one or more **protection circuits**, varying from resettable tweeter protection circuits to conventional fuses.

- On some cabinets, you can **adjust the level** of one (the tweeter) or more speakers.

Bi-amping

Powered speakers often feature bi-amping, with power ratings varying from 100 watts for the lows and 30 for the highs, to 400 or

more for the lows and a 100 for the highs. Some companies offer tri-amping systems with a total power rating of 900 watts or more. Powered subwoofers have power ratings up to 750 watts.

Class H

To save weight and reduce power consumption and heat, many powered speakers use class H amps. Contrary to traditional class A/B or class A amps (see page 117), a class H amp only draws power when it really needs it. If not, it runs almost idle. The resulting lower operating temperature also reduces the need for heat sinks or fans.

TIP

Features: powered speakers

Some powered speakers have sufficient features to double as a keyboard amp or an acoustic amp, with a number of line and/ or microphone inputs, multiple channels and level controls (so you can use it as mixer too), a headphone jack, effects loop, and tone controls. Other features you may come across are a loudness switch (boosts highs and lows for low volume situations), a DI with a ground lift switch or a line output, and a variable crossover filter for subwoofers (see page 133).

Digital processors

A growing number of powered speakers also have integrated digital processors providing anything from a noise gate and multi-band compressor/limiter to a delay (to eliminate phase differences between various speakers), a fully automatic feedback filter, a hum canceller, or an adjustable low-pass filter for a more refined bass sound. *Tip:* Some models can be hooked up to a PC, allowing you to fine-tune the parameters of these and other built-in effects and controls.

Presets

Rather than fully adjustable tone controls, some models have a number of presets, offering less flexibility but easier control. Presets may include settings such as:

167

- **outdoor use** (cuts lows and highs to aid projection);

- **CD player** (for a less aggressive, hi-fi type of sound);

- **wall mounted** speakers;

- mic plug and play (for **speech**; cuts highs and lows);

- '**flat**' (use a separate equalizer if required).

These presets often incorporate an integrated compressor or feedback filters.

Rear panel of a powered monitor.

MONITORS

If you don't have separate monitor speakers, you'll need to position the PA speakers behind the band in order to be able to hear yourself play. This increases the risk of feedback problems, however, and it exposes you to the highest sound level in the room — close to the speakers. This can be both tiring (but inspiring for some) and harmful to your hearing. Using monitors allows you to crank the volume up out front, while adjusting the monitor levels so that you can hear yourself play at a reasonable volume.

168

Four types

Basically, there are four types of monitors. The wedge monitor, designed not to block the audience's view; the *sidefill*, a full-range speaker that allows for a stage mix of the full band; personal monitors that are small enough to be mounted on a microphone stand; and *in-ear monitors*. The latter two solutions reduce the risk of feedback and they reduce the onstage sound level, especially in the case of in-ear monitors. This doesn't work for everyone: Especially in heavier styles of music, musicians often want to feel the beat as much as hear it.

In-ear monitors

One of the main advantages of in-ear monitors is that they don't have to compete with the sound level of other types of onstage monitors. They're in your ears, effectively blocking out the stage sound, so you can use them at a very moderate volume. This helps prevent hearing loss and it helps singers save their voice as they don't have to strain to sing over the band. That makes singing in tune easier too. A disadvantage is that they can hinder communication with your fellow band members and your audience, and they're quite expensive: Wireless in-ear systems start around six hundred dollars.

TESTING SOUND SYSTEMS: TIPS

- In judging sound systems, most of the **general tips** from Chapter 6 apply.

- **Sound dispersion** (see page 73) is an important topic for FOH speakers. The sound should be the same whether you're in the middle, between the speakers, or to the side of the hall.

- Likewise, the sound should **project**: It shouldn't change (other than getting gradually softer) when you move away from the speakers.

- Sound systems are designed to produce a **natural** sound. They need to address much larger audiences and require much larger

169

sound pressure levels and a more aggressive type of sound than a typical home stereo system.

• Bring along **the musicians who are going to use the system**, and let them listen to the system or try it out if possible. Always check how the singer sounds through the system! You can also compare systems using a CD that you know very well. Preferably use a recording of your own band, so you know how you want things to sound.

• Music stores are **rarely large enough** to really try out sound systems, so you may need to rely on the salesperson's information even more than with other amps.

12

Effects

The similarities between effects and spices are striking: They can be used to add a subtle touch or to completely disguise the original, they can be used one at a time — and be very effective — or in any number of combinations, and there's an unlimited variety available. This chapter covers the main effects, and how to combine and test them.

Electric guitarists are the biggest effects users, followed by keyboardists and electric bassists. The number of dedicated effects devices for acoustic guitarists and vocalists (see page 201) is more limited, and there are even fewer effects that were specifically designed for other musicians, but there's still plenty to experiment with.

Dedicated effects

That said, some effects just don't work for certain instruments. A chorus sounds great on an electric piano, but it can make an acoustic piano sound as if it's terribly out of tune. And while many guitar effects can be used for bass guitar too, there are just as many others that thin out the bass sound, reducing its fundamentals, punch, and definition, so it gets lost in the mix.

Categories?

Classifying effects is tricky business, as there are many effects that fit more than one category. The following four categories aren't perfect either, but they do help to get a basic grip on the many types of effects there are. This chapter covers effects in this order:

- **Ambience**: Reverb emulates the sound reflections in a room, while a delay basically simulates an echo (but there's a lot more you can do with it!).

- **Tone**: Effects that alter the tone by adding a copy of the signal to it (such as chorus or flanger), by adding harmonics (distortion, overdrive), by cutting or boosting frequencies (EQ), or by changing the harmonic content (wah-wahs and synths).

- **Pitch**: Effects that change the pitch or add pitches to the original tone, such as pitch shifters and detuners.

- **Dynamics**: Effects that change volume levels, such as tremolo or compression, which boosts soft signals and cuts peaks.

(Modeling) preamps

As a preamp's controls (volume, EQ) can also be used to color and change the sound, preamps can be considered effects as well — especially modeling preamps. There's a separate section on these units on pages 208–209.

172

Parameters

Each effect can be fine-tuned with one or more controls that set the effect's parameters. Here are some common *parameters* on various types of effects:

- The **level** knob is used to set the effect level. This control can be used so that the overall volume level doesn't change when switching the effect on or off, or to add more or less of the effect to the sound.

- A **tone** control cuts or boosts highs and/or lows.

- The **rate** or **speed** control sets the 'speed' of the effect. On a vibrato, for example, it controls how fast the pitch goes up and down.

- The **depth** or **intensity** knob on the same vibrato is used to set *how much* the pitch goes up or down.

- A **delay** or **predelay** control can be used to postpone the effect a little: First you hear the dry sound, then the effect sets in. Note that 'delay' is also the name of a certain type of effect (see pages 175–178).

- With a **threshold** or **sensitivity** knob you can set an effect so that it only works above or below a certain level.

Tips

There are many more parameters, some of which you will come across on the following pages. Tips:

- *Some parameters come under **different names**, as explained above, just like different parameters may share the same name.*

- *The **number of adjustable parameters** of an effect depends on the effect itself, as well as on the device.*

- *Effects that are built into amplifiers usually offer **less control** than effects in dedicated units.*

173

- A **wet/dry** or **mix** control sets the balance between the effected (wet) and the uneffected (dry) signal. On some units, this control is called **blend** or **level**.

- A **feedback**, **resonance**, or **regeneration** control allows you to feed more or less of the output signal back to the effect input, so it will be processed again.

AMBIENCE: REVERB

Singing in the bathroom sounds so good because of the way the sound reflects back at you. This is known as reverb, short for *reverberation*. Without reverb, amplified musical instruments tend to sound very 'dry.'

Early and late
Clapping your hands in a room with your eyes closed can give you a general idea of the room: The first reflections or *early reflections* you hear tell you whether you're in a cathedral (large, with hard walls) or a living room (small, with rugs and curtains), for example. The early reflections are followed by the diffuse *late reflections*, all arriving at different rates.

Types of reverb
There are many different types of reverb.

- Most guitar amps have a **spring reverb** (see page 119).

TIPCODE

Tipcode AMPS-013
This Tipcode demonstrates eight different types of reverb, including room, church, and plate.

174

- The bright, dense sound of a **plate reverb** is commonly found in digital units only. Real plate reverbs (a thin metal plate in a soundproof enclosure) are big and expensive, so they're basically just used in studios.

- Digital reverbs usually offer a number of reverb types, most of which are **named after the room they emulate**: room, stage, auditorium, concert hall, or church, for example. Most digital reverbs also offer simulations of spring and/or plate reverbs.

Reverb parameters

The predelay controls when the reverb kicks in, while reverb time (or decay, or decay time) controls how long you will hear the reverb. A longer time emulates a larger room, or a room with harder floors or walls (or a smaller audience). Use shorter reverb times as the music is denser and the tempo is higher, and save long reverbs for ballads. You may be able to adjust the tone as well, making for a warmer or a brighter reverb.

Reverb variations

A *gated reverb* is mainly used for drums, the gate suddenly closing and thus cutting off the *reverb tail* after a set time, or at a set sound level. This type of reverb was made popular by drummer and singer Phil Collins. A *reverse reverb* gets louder, rather than softer.

Listen

Reverbs are usually meant to be lush and warm or crisp and clean, rather than undefined or metallic. That said, some musicians do prefer that metallic type of sound, using their reverb units as a true 'effect' rather than to recreate the acoustics of a room for a more natural type of sound.

TIP

DELAY

Delay is also known as *echo*: It makes a copy of the sound and

175

repeats it once or a number of times, with the echoes gradually getting softer.

Delay parameters

The delay time controls the time between the original signal and the echoes. Delay times may range from less than 20 ms (milliseconds) to 5,000 ms (5 seconds) or more. Feedback sets the number of repeats. A mix or level control sets the volume of the repeats.

TIPCODE

Tipcode AMPS-014
This Tipcode demonstrates various types of delay on guitar.

Delay types

A delay is a very versatile effect. A few examples:

- **Doubling**: With a very short delay time (up to around 40 ms), the sound is doubled, rather than producing a distinct echo. Doubling makes for a 'fatter' sound.

- A delay time of 40 to 140 ms gives you the famous rockabilly **slapback** effect.

- With long delay times, you can **play along with yourself**, either creating harmonies by playing long notes, or playing loops of multiple shorter notes, for example.

- With a (stereo) **pingpong** delay or **panning** delay the sound moves from left to right and back.

- A **reverse delay** reverses the sound, echoing what you play backwards.

- A **multitap delay** provides you with multiple delay times, allowing for a very broad sound and complex, rhythmic delay patterns.

176

- Just like a reverb, a delay can be used for a more **ambient sound**, reducing the number of repeats and very subtly adding it to the mix.

Double your tempo

A delay can make you sound as if you're playing twice as fast. Set the delay time so that it produces an echo in-between the notes you're playing. If you play eighth notes at 110 BPM, and you want to hear sixteenths, you divide 60 (seconds per minute) by 110 (BPM), and divide the result (540 ms, equaling one quarter note) by 4 (to produce sixteenths), so you set the delay time at 540÷4=135 ms. This requires a digital delay that allows you to set exact delay times.

TIP

Delay time

If you use a delay to produce distinct echoes, the delay time should match the tempo of the song. A *tap tempo* control is a great help for that purpose: Simply tap or stomp the tap button a couple of times in time with the music. Most delay pedals, and some other effects, feature this control.

Echo

The original delay units consisted of a tape loop, a recording head and a playback head. Delay time was set by moving the playback head back and forth. Each following repeat sounded noticeably duller and softer than the previous one, just like a natural echo. When emulating this effect, many companies use the term 'echo' rather than 'delay.' The warbling effect of a worn-out tape can be digitally imitated as well.

Ducking and spillover

Some units can reduce delay and reverb levels when a lot of notes are being played. This is known as *ducking*. Delay and reverb *spillover* makes sure the tails of these effects continue when you switch to another preset, rather than cutting them off.

CHAPTER 12

A delay unit with a tape imprint. The black fader below this 'tape' emulates the original playback head. Moving it to the right increases the delay time.

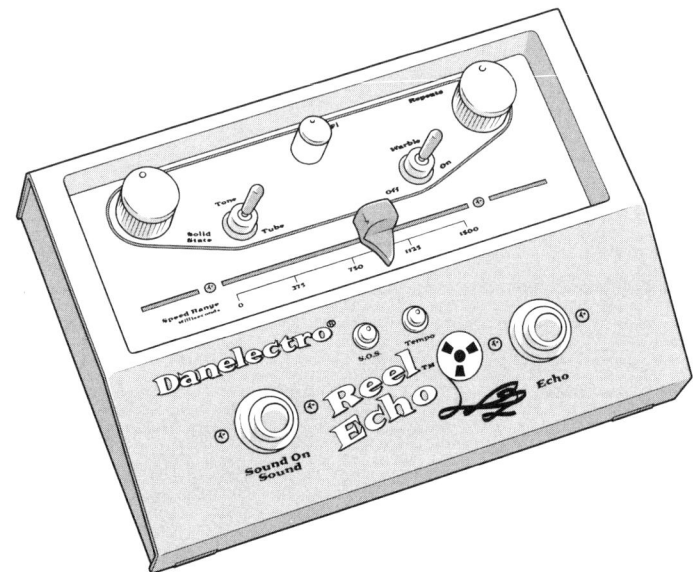

Ducking, part II

When talking delays, ducking refers to suppressing the effect as long as you're still playing: The effect starts when you play your final note. Ducking is also used to describe the effect of having one audio signal suppress the other: a DJ can suppress the music by simply talking into the microphone.

TONE

The first four effects covered in the 'tone' category also belong to the group of *modulation effects*: chorus, flanger, phaser, and ring modulation. (This group also includes other effects that 'modulate' the sound, such as vibrato and tremolo.)

Time-based effects

Chorus, flanger, and phaser are also labeled as *delay effects* or *time-based effects*, just like a delay, indeed: They combine a delayed copy of the signal with the original signal.

178

TIPBOOK AMPLIFIERS AND EFFECTS

Delay times

A chorus uses delay times of approximately 30 to 50 ms. For the swirling sound of a flanger, the delay time is usually between 10 and 20 ms, and the woosh of a phaser requires some 5 to 10 ms.

CHORUS

For a *chorus*, a copy of the signal is added to the original with varying delay times. This variation causes pitch differences, which make the sound swirl or shimmer, as if you hear two instruments simultaneously — or as if you hear a 'chorus' instead of a single voice.

Thick, warm, big

When used subtly, a chorus makes the sound warmer, thicker, richer, creamier, wider, fatter, or bigger than it really is. It is a great and widely popular 'spatial' effect for fretless bass guitars, strings, electric pianos, jazzy guitar chords and some singing voices, for example, but it can also be used to warm up a distorted guitar sound. Cranking the rate and depth controls turns a chorus into a 'bubbling' effect.

More

A *multi-voice chorus* supplies you with two, four, or more copies of the original. A good chorus has a lush, full, thick, yet clear sound with sufficient depth.

FLANGER

A *flanger* is a swirling effect that can range from a modest swoosh or a chiming, metallic effect to the sound of an angry jet plane mixed in with the original music. Flangers sound great on fretless basses, some electric guitar sounds, and electric pianos. Standard

179

controls are depth, rate and delay. The flanger effect can be enhanced by using feedback or resonance.

Flanging

The term 'flanging' was presumably coined when two tape machines were simultaneously playing the same song for a delay effect at a Beatles recording, and someone slowed down the pitch of one of the tapes by holding its tape reel by the flange.

PHASER

A phaser or phase shifter is a third variation. The copied signal is shifted in and out of phase with the original at varying delay times. As a result, harmonics are boosted or reduced, creating a sound that can best be described as twisting or, again, as sweeping or swirling. The effect is also compared to the spinning sound of a rotary speaker (see page 181; Leslie), while others describe a phaser as a mild, drier version of a flanger.

Rich

Phasing is applied to rhythm guitar patterns in reggae, funk, fusion, and other styles, as well as to electric pianos. Both flangers and phasers usually work best on harmonically rich sounds: If there are few harmonics to begin with, these effects might thin out the sound.

TIPCODE

Tipcode AMPS-015
Play this Tipcode for a brief demonstration of the effects of a phaser, a chorus, and a flanger on the sound of a solidbody electric.

180

RING MODULATION

A *ring modulator* or *balanced modulator* generates frequencies that are not harmonically related to the original, resulting in weird, unexpected and often non-matching frequencies that remind most people of sci-fi movies and robots, or cracked church bells. The effect was mainly used on electric pianos in the early days of fusion in the 1970s.

LESLIE

Yet another modulation effect stems from the classic *Leslie speaker*, a powered enclosure that features a rotating horn for the treble range and a rotating drum that swirls and projects the woofer's low frequencies. This phaser related effect works great on tone wheel (Hammond) organs and some type of guitar sounds. Leslie is a registered trademark, so digital and analog simulations of this effect are given names such as *rotor* or *spatial sound*.

DISTORTION

The widely used distortion effect stems from overdriving low-powered amps (see page 17). It is mainly used for guitars, but electric pianos, tone wheel organs and other instruments may benefit from it as well. Distortion effects are available in various 'degrees,' from a relatively mild overdrive to a high-gain distortion. Next to the common controls, distortion effects often have parameters with colorful names such as punch, muscle, or fat.

Fuzz
The *fuzz box* is one of the earliest guitar effects, dating back to the early 1960s. The sound is best described by its name: fuzz.

181

Overdrive

Another descriptive name: An *overdrive* emulates a cranked-up, overdriven tube amp. The sound of an overdrive is often described as warm, smooth, and creamy, reminiscent of power tube distortion (see page 115). Blues players, among others, use an overdrive rather than a harder type of distortion. If you set an overdrive's gain rather low and its level control high, you can use it as a *booster*, increasing your volume for a solo. Dedicated boosters are available too.

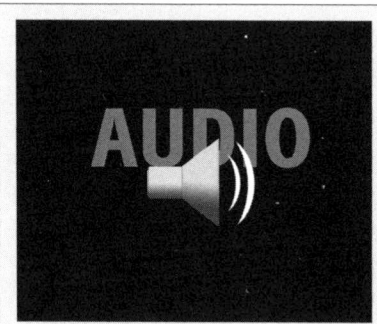

Tipcode AMPS-016
Listen to this Tipcode to hear the effects of an overdrive pedal.

Distortion

Compared to an overdrive, a genuine *distortion pedal* produces a harder, more metallic, high-gain type of distortion.

Tipcode AMPS-017
Compared to the overdrive in Tipcode Amps–016, a metal distortion pedal makes for a very different sound.

Tube distortion

If you have a solid-state amp and you want real tube distortion, you can get preamps that feature one or more preamp tubes, or an

182

extremely low-wattage, easily overdriven power amp. Hook up the preamp to your effect return jack, or to another power amp, and go.

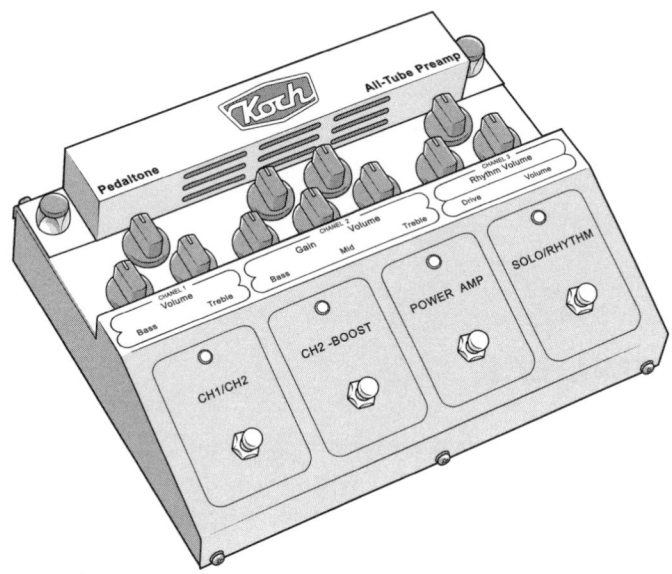

A multi-channel tube preamp with independent tone controls and a built-in 0.5 watt tube power amp.

Bass distortion

Grunge and metal bass players use distortion too. In order to maintain definition and punch, bass distortion pedals usually allow you to create a mix of the original sound and the distorted signal.

Listen

What distortions can sound like is on page 120 (lead channel). Most guitarists don't want the effect to cover up the differences between their guitars, or between their bridge and neck pickups, and they often *do* want them to be dynamic, like a tube amp, cleaning up when you play softer, yet without losing body.

Tip

Distortion pedals are designed to be used with an amp's clean channel. If you're feeding a distorted signal to the overdrive channel, things can get out of hand — but if you play nu-metal, this may be the sound you're looking for.

183

TIP

Listen

When judging distortion effects, try to find out if you can still hear the difference between your bridge and neck pickups, and check how your instrument responds to your playing. For example, the sound may become a bit cleaner as you play softer, but doesn't it also become kind of thin, lacking body?

EQUALIZER

Stand-alone equalizers come in stompbox and rack-mountable formats. Why would you buy a separate equalizer if your amp has tone controls already, and possibly your instrument too?

- If you use **two instruments**, you can adjust your amp's tone controls for one, and the equalizer's controls for the other instrument, so you don't need to compromise.

- Likewise, you can use the equalizer unit to set the tone of your clean or overdrive channel if your guitar amp doesn't have **independent tone controls**.

- A dedicated equalizer often has a **more detailed and stronger effect** on the tone, having more bands, and cutting or boosting more decibels. Getting a powerful equalizer is almost like adding an extra channel to your amp.

- An equalizer can double as a **preset volume pedal**: Boost all frequencies the same amount and the sound will get louder (or gain will be boosted) as you switch the effect on.

- Multi-band equalizers can be used as a **feedback filter** too (see page 247).

184

WAH-WAH

A *wah-wah* is an extreme, foot-operated tone control that clearly
sounds its name as you rock the effect's expression pedal up and
down. This makes a filter sweep through a range of frequencies,
usually between 400Hz and 2kHz. You can imitate the effect to
some extent by rhythmically rotating the frequency control of
a parametric (sweepable) equalizer. Harmonica players make
wah-type effects by moving their hand to and away from the
instrument, and brass players use a hat or a special mute in front
of their instrument's bell.

Parameters
Apart from a number of adjustable parameters (including the
width of the frequency band or Q), you can, of course, vary the
speed at which you move the pedal up and down — usually in
time with the music, but moving it up and down very slowly
creates flanger-like effects.

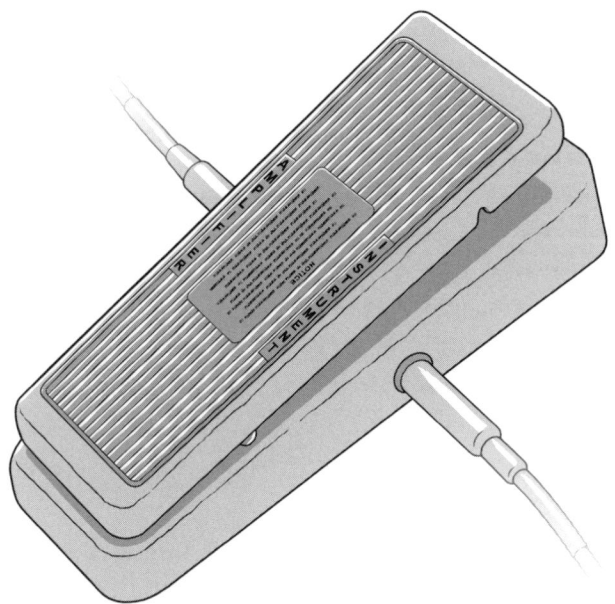

A wah-wah
pedal
(Dunlop)

Tipcode AMPS-018
A wah-wah sounds very much like its name, as this Tipcode clearly demonstrates.

Wacka-wacka

Every company has its own ideas of how a wah-wah should sound. Their various models sport sounds ranging from funky wacka-wacka or wocka-wocka effects to smooth, fat, round, or aggressive wah sounds. Variations on the theme include effects that produce 'yah-yah' and other vowel-like sounds.

Notched

Rather than moving the pedal up and down, you can also leave it in a notched position for a special nasal or vocal type of sound. This doesn't work on spring-loaded pedals, which move up as soon as you take your foot off them.

TIP

Optical or analog

Some pedals use an optical sensor rather than an analog potentiometer to sense the pedal movement. Potentiometers wear down eventually; replacing them is a costly affair. Optical sensors don't have this problem, and they respond very quickly. Additional tip: When judging wah-wah pedals, listen whether the pedal affects the sound in its off position. It shouldn't.

Auto-wah

An *auto-wah* has no pedal to control the wah. Instead, the effect is controlled by your dynamics, mimicking the effect of moving the pedal down at each note or chord you play, and letting go as the

186

sound gets softer. In other words, the effect follows the *envelope* (attack, decay, sustain, release) of your notes. The louder you play, the further the virtual pedal will be depressed at the attack. Some auto-wahs allow you to set the effect's response or starting level with a sensitivity or threshold control. A peak control adjusts the width of the wah's filter. The speed at which the 'pedal' moves up and down may be adjustable as well.

Tipcode AMPS-019
Here, you'll hear the effect of an auto-wah on a funky piano sound.

TIPCODE

Envelope
An auto-wah is an *envelope follower* or *envelope filter*. It follows and responds to the envelope of every note you play. Envelope filters can be applied to other effects as well.

Combinations
Wah-wah pedals are often combined with a volume pedal, and others include fuzz or overdrive effects.

AND FURTHERMORE

There are many other effects that fit this category. *Synthesizer effects*, for example, that add a synthesized tone to the original signal. They're usually used to create fat, classic analog synth sounds on basses or guitars. With a *talk box*, you speak or sing into a microphone as you play, and your instrument will sound like it can talk. A *vocoder* can mix your voice and an instrument signal to create a robotic kind of sound. In some styles of dance music,

187

the sound is supposed to be *lo-fi*, and there are special effects that make your instrument sound like an old vinyl record.

PITCH

On guitars, violins and other stringed instruments, you create vibrato by rocking the finger that stops the string back and forth. This makes the pitch go slightly up and down. The main parameters, both for an 'acoustic' or an electronic vibrato, are rate and depth (how fast and much the pitch goes up and down).

Pitch shifting
A *pitch shifter* adds one or more notes to the note you play or sing, turning your single notes into chords or harmonies, or turning a single singer into a backing vocal group. The added notes can be higher and/or lower than the notes you play or sing.

Intelligent pitch shifters
Intelligent pitch shifters such as the Harmonizer also take the key signature of the song into account, so they know whether to add either a major or a minor third, for example.

Octavers
An *octaver* is a very basic type of pitch shifter. It adds a single note one or two octaves below the original note.

TIPCODE

Tipcode AMPS-020
This Tipcode demonstrates some of the things you can do with a pitch shifter and an octaver.

Range and tracking

When checking out pitch shifters, find out how large their range is (an octave up and down, or more) and listen whether they follow you accurately at any speed and pitch (*tracking*).

Whammy

A *Whammy pedal* (a Digitech exclusive) simulates the effect of an electric guitar's whammy bar or tremolo arm, which tightens or loosens the tension on all the strings simultaneously, bending the overall pitch up or down. An electronic whammy pedal allows you to do this over a much larger range, up to one or two octaves, adjustable in half steps (e.g., 2, 5, 7, 12 or 24 half steps; 24 half steps equaling two octaves). A pitch shifter with a continuous controller pedal (see page 204) can be used for a similar effect.

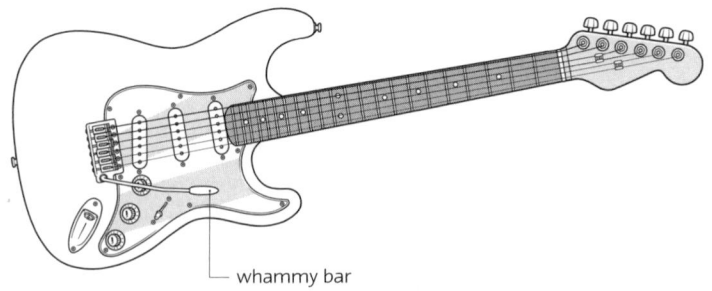

whammy bar

Depressing the whammy bar (often incorrectly labeled tremolo bar) lowers the strings' pitch for 'divebombing' and other effects.

Detuner

A *detuner* makes for a thicker sound by adding a slightly detuned signal to the original. The word D-tuner sounds the same, but it refers to a device that allows you to quickly drop the pitch of a guitar or bass string, usually tuning low E to low D.

Pitch correction

A very special type of effect is used to make vocalists sing in tune, correcting their pitch in real time. Pitch correction is used both in the studio and live, and it's a common feature in multi-effects units for singers (see page 201).

189

DYNAMIC EFFECTS

The most basic dynamic effect is the volume pedal. Depressing it makes you sound louder, so guitarists can make their notes come up slowly, just like violinists or horn players (you can do so with your guitar's volume control too, but a pedal is much easier).

Tremolo
A tremolo rhythmically varies the volume level. This is called *amplitude modulation*. It's often confused with vibrato, but vibrato is *pitch* modulation.

Pingpong
A *pingpong* effect (a.k.a. *panner*) makes the sound go from left to right and back, as if someone keeps turning the balance control back and forth.

Delayed attack
A *delayed attack* sounds as if you're turning the volume control up from zero for each new note. You can also do this with a volume pedal, but this *slow gear* or *auto swell* effect makes things a lot easier.

COMPRESSORS AND EXPANDERS

Some effects seem to be designed to control the sound rather than to enhance it, but they can be used both ways. Compressors and expanders, for example!

Compressor
A *compressor* suppresses loud signals, literally compressing them. This reduces the dynamics of your performance. Compressions works great for singers, among others: Reducing unintentional dynamic differences improves intelligibility.

Threshold and ratio

If a sound is louder than the value set with the compressor's *threshold* control, it is reduced to a certain degree, referred to as the *ratio*. For example: If the ratio is set at 3:1, the peak of a signal that crosses the threshold by 3dB will be reduced to 1dB. If the threshold is crossed by 6dB at that same setting, the output will go up 2dB, and so on.

More parameters

Some of the other compressor parameters are attack, release, hard or soft knee, and output or gain.

- The **attack** controls if compression kicks in right away, or a bit later. If the attack is set too short, you won't hear the stick attack on a drum anymore, for example.

- The **release** controls when the compressor stops working as the signal gets lower than the threshold. Release and attack times can commonly be set from a few milliseconds to a few seconds.

- Set at **soft knee**, compression sets in gradually, making the effect less obtrusive and more natural. The hard knee compression is more accurate or aggressive, and often more effective. A hard knee setting also helps if you lose punch at low ratio settings.

- An **output control** or **gain control** can be used to boost the reduced overall signal.

Punch and sustain

A compressor can be used to add punch or sustain to your sound. Want more punch? Bring the threshold back to −20 or −25dB, go for a 4:1 ratio, a 100 ms attack and a 1,000 ms (1 second) release. More sustain for your guitar solo? Try a −30 to −35dB threshold, a 10:1 ratio, a short attack (5 ms) and a long release (3 to 4 seconds). This can be very effective, which explains why guitarists' compressors are often labeled *compressor/sustainer*: The long release time makes the sound go on. The signal is boosted as soon as it falls under the threshold.

Squashing, pumping and breathing

Learning to set a compressor takes time. If you don't do it right,

191

the effect may squash the sound, or you may hear pumping (release time too short) or breathing (release time too long) sounds. Extensive control over the parameters allows for an inconspicuous use of the effect, which is what you're usually after. Of course, a good and well-adjusted compressor does retain the typical differences in timbre between playing of singing louder of softer, but it balances out the dynamics.

Bass amps

Bass amps often have a built-in compressor. It may be either non-adjustable or linked to the gain control: The higher the setting, the more compression you will get. More advanced models may offer more control over the parameters.

LIMITER

A *limiter* doesn't allow the sound to cross a certain threshold at all: It's like a ceiling, and the music can't get any louder than that. Some limiters have a very audible threshold level, where you can hear the sound getting squashed against the ceiling. Limiters help protect your amp and speakers. Disabling the limiter increases the risk of damaging your equipment, but it may breathe more air into your sound.

Compressor/limiter

If you set a compressor to a high ratio of 20:1 or ∞:1 (infinity to one) it turns into a limiter.

TIP

Too low

Warning: Setting the threshold level too low may result in distortion and subsequent damage to your speaker(s).

EXPANDERS AND NOISE GATES

An *expander* is the opposite of a compressor: It attenuates quiet signals and boosts louder signals. In other words, it increases the dynamics of a signal, rather than reducing them. Just like a compressor, an expander may let you control threshold and ratio (or rate) as well as attack, release, hold, or decay times. And like a compressor, an expander can be used to enhance the sound too, for instance to make percussive sounds punchier and tighter.

Ranges
The threshold range may be adjustable from -50 or -60 to 0dB. Similar to a compressor, attack times usually range from 1 ms or less to 300 ms or more, and release times may be as long as several seconds.

Noise gate
If you set the ratio very high (1:100), the expander acts as a *noise gate*: a gate that blocks noise, reducing the gain to zero. Amps and electric or electronic instruments tend to produce noise that is not audible when you play, but it is when the music stops. A noise gate can be used to cut out other types of noise too, from the buzzing of a snare drum to the breathing of a singer, or sounds from other sources that bleed into open microphones. If the trumpet plays a solo and no one else is playing, you don't want any of the other mics to pick up the trumpet sound.

Listen
Noise gates and expanders should be set so they don't chop off the sound (unless you want them to), and they shouldn't influence the timbre of the instrument.

EVEN MORE EFFECTS

- There are many more effects that fit one or more of the above categories, often featuring more or less descriptive **fantasy**

193

names, from Black Cat Moan and Jackhammer to Crybaby (wah-wah), Small Stone (phaser), Flerb (flanger and reverb), Ice Box (chorus), Tuna Melt (tremolo), or Holier Grail (reverb).

- Various effects are available in **stereo**, adding depth and space to the sound. Some examples would be chorus, phaser, reverb, and delay. One use of a stereo effect such as a chorus is to have split wet and dry signals that alternate from left to right.

- Noise gates and expanders are effects that are usually used in studio and PA situations. Other typical **studio effects**, such as enhancers or vitalizers, help breathe life into studio recordings.

TESTING EFFECTS: TIPS

- Many of the tips listed in **Chapter 6, Testing Amps**, apply to effects too.

- Try effects with the **instrument, the amp, and the other effects** you're going to use them with. An effect that sounds great with one amp may be disappointing with another. *Tip:* With some effects, you'll want to adjust your amp's EQ.

- Try effects **both** in the amp's loop, if available, and in the instrument input.

- Take your time to check out the effect's **parameters**. More expensive units commonly offer more controls, and their knobs are likely to provide both subtle control and more extreme settings.

- Know **what to expect** from an effect. A pitch shifter may be great for solos, but it will usually make a mess of chords.

- Factory presets of multi-effects units are commonly **designed to impress**. They may sound great in the store or at home, but they're often too much of a good thing when used in a band setting. If you don't like the factory presets to begin with, you'd probably better look for a different unit altogether.

194

- When checking out multi-effects units, listen for **dropouts between preset changes**.

- **Analog stompboxes** usually produce more noise and hiss while digital effects tend to sound clean and crisp. They're also much more predictable — or less adventurous. Which is better? That totally depends on what you like. Check out pages 67–68!

- Listen carefully for **unwanted side effects**: Effects can add noise, and they can make your instrument sound thinner (eating up bass), duller (rolling off highs) or less direct. They can also compress your sound, reducing your dynamics.

- Some companies are known for their **aggressive effects sounds**, while others produce similar effects with a sweeter, less-obtrusive character. Finding out which brand you generally like best can make you search a lot easier, and combining effects from various manufacturers may result in a very colorful effects setup.

CONNECTING EFFECTS

Effects can be connected between the instrument and the amp or in the effects loop, if available. Using the loop might help produce a cleaner, more natural sound, and you may be able to switch all the effects in the loop on and off simultaneously, or control their overall level with a single level control on the amp.

TIP

For guitarists

For guitarists, the difference between using the amp's instrument input or the effects loop is larger than for most other musicians: It is the difference between applying an effect pre-distortion (when using the instrument input) or post-distortion (using the loop).

195

Long signal path

If you want to have your effects unit nearby and your amp is at a distance, using the effects loop requires long cable runs (one from the amp to the unit, and one back to the amp), which may generate noise.

Compressors and overdrives

Many guitarists run their compressor and distortion or overdrive pedals into the guitar input, and their ambient and modulation effects in the loop. This makes the latter effects sound more dominant as they affect the distorted signal, rather than the other way around. However, putting them in the loop can make some effects sound noisy.

Wahs and choruses

Wahs are usually plugged into the amp's input. Using the loop makes them sound fatter, which may or may not be what you want. The best tip is to experiment. Some players use their chorus in the loop, others prefer to use it before the preamp — and some do it one way today, and the other way tomorrow.

Multiple effects

When you're using multiple effects, you will find that the order of the effects in the chain is essential to the resulting sound. Most multi-effects units have controls to set the effects order, and with stompboxes it is simply a matter of connecting effect outputs to effect inputs in any order you like.

TIP

Too many

If you use too many effects simultaneously, your sound will get messy and much of what you play can get lost. The same goes for extreme effect settings. Effect combinations and settings that sound great at home can easily be too much in a band situation. Less is often more: less effects, and lower settings of depth, rate, and other controls.

Experiment

Determining the order of the effects in the chain is largely a matter of experimenting. The following are some things to consider. One tip in advance: If you want to use a tuner in the chain, it always comes between your instrument and the first effect. Tuners don't like effected signals (especially pitch-change effects such as vibrato!).

Anticipate

The better you understand what effects do, the easier it is to anticipate what certain effect orders will do. An example? Distortion means compression, reducing dynamic differences. That means that an auto-wah, which responds to dynamic differences, should come before distortion (and before a compressor, of course).

Tipcode AMPS-021 & 022
Play Tipcode Amps–021 to hear a combination of a compressor, a chorus, and a delay. In Tipcode Amps–022, four effects are used simultaneously: compressor, overdrive, chorus, and reverb.

TIPCODE

The chain

The following order of effects may be a helpful guideline, derived from the experiences of various musicians. The effects in parentheses can be used in a variety of positions in the chain. Opinions differ on which position is best.

Octaver ➔ auto-wah ➔ compressor ➔ wah ➔ (delay) ➔ distortion ➔ fuzz ➔ overdrive ➔ (gate) ➔ EQ ➔ (flanger/phaser/chorus) ➔ volume ➔ delay ➔ (fl/ph/ch) ➔ (gate) ➔ reverb > (gate).

Series or parallel

Connecting stompbox effects output to input means they're

197

connected *in series*. This means that the first effect always influences the second, and so on. You can also connect them *in parallel* so they both influence the signal, but they don't affect each other. To do so, use a box with a single input (for your guitar) and two more outputs (that go to the effects) and a second box with multiple inputs (from the effects) and a single output (to the amp). A chorus and a delay, for example, will sound different in series than in parallel. Some multi-effects units enable you to use effects both in series or in parallel.

More chains

A *loop switcher* or *effect looper* is a small box that allows you to use two (or more) different effects chains or loops, and you can switch all the effects in a chain on or off with a single button.

Mud

If your amp's signal goes to the PA, the sound engineer may not want you to use any heavy time-based effects at all. To keep your sound from getting muddied up by the time it reaches the sound system's speakers, these effects should be close to the end of the chain: the mixing console's effects loop, in this case.

13

Effect Equipment

Effects come in various guises, built into amps, stompboxes, floor-based or desktop multi-effects units, rack-mountable units, and other housings. Here's a closer look at what's available and what to pay special attention to.

One of the main differences between built-in effects and dedicated effects units is the control you have over the effects' parameters. For built-in effects, this is usually quite limited.

Home, studio, stage

Stompboxes or compact effects are mainly designed for onstage use, as are floor-based multi-effects units. At home, a desktop model often makes more sense, while rack-mountable units that usually feature fewer effects, more parameters, and a higher quality and price, are typically used in studios and large PA systems. To suit everyone's needs, some companies have multi-effects or modeling preamps that come in all of these versions: floor-based, desktop and rack-mountable. *Tip:* Some desktop units are stand-mountable, which makes them easier to use onstage.

Workstations

If you really can't choose, you can get an all-in-one unit that can be used for live gigs, home studio sessions, practicing, and even composing. These so-called *workstations* feature a large number of effects as well as amp, cabinet, and stompbox models. They often sport user-programmable presets and artist-designed factory presets, as well as a drum machine, a bass machine, and an MP3 player to jam along with. Some come with a microphone preamp (with phantom power) for singing players or talkbox effects, a multitrack recorder that helps you compose songs, and USB or Firewire and MIDI ports that allow you to store or edit presets and songs on your computer (or to download patches from other users).

Non-guitarists

Many non-guitarists use guitar effects — either because they like to experiment, or because that used to be all that was available. But the number of dedicated bass effects has been growing steadily, and some companies make special effects devices for acoustic guitarists and singers. The latter are also used by horn players, flutists, and other acoustic musicians.

Acoustic guitarists

Some features that you may find on multi-effects units for acoustic guitarists are microphone modeling (see page 25), EQ, feedback reduction, a tuner, a headphone output allowing you to silently use the built-in effects (chorus, delay, reverb, and so on) and an expression pedal or expression pedal input.

Vocals and horns

Singers can use the same effects as horn players, including chorus, flanger, phaser, tremolo, detune, delay, and wah. Compression is a very useful effect too. Pitch shifting allows singers to sound like a quartet or a choir all by themselves, a *de-esser* reduces the sizzling effects of the vowel 's' and other sibilant sounds, and pitch correction (see page 189) makes vocalists sound in tune. These effects are also available in separate units.

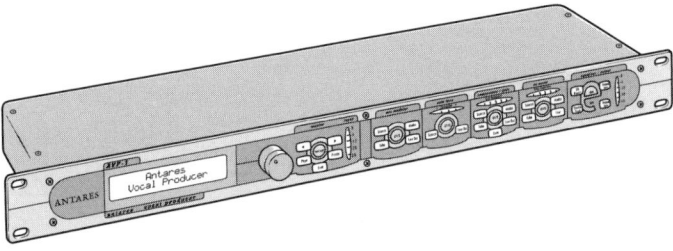

Special rack effects unit for singers, featuring (Auto-Tune) pitch correction, microphone modeling, a compressor, a de-esser, and more (Antares).

Reverb and compression

Reverb and compression are two other effects that vocalists and horn players often benefit from. It's often worthwhile to invest in these tools if you perform without a professional sound system and an engineer most of the time.

Tubes

Just like amps, effects units may use one or more tubes to make for a warmer, smoother, or creamier sound.

STOMPBOXES

Effects pedals or stompboxes are the most basic types of effects units. A limited number of dedicated controls makes them easy to

201

use. Some units feature recessed or spring-loaded controls to avoid damage or unintended adjustments by stomping feet. The housing can be plastic or metal, the latter promoting a longer life expectancy. (There are companies that offer a no less than six-year warranty.) Besides the common rectangular 'boxes', there are stompboxes in a variety of spectacular colors and designs, including car-like models and foot-shaped rockers. Most companies use a different color for each type of effect, so they're easily identifiable.

Stompboxes have either a switch button (or two) or a pad. The pedal on the left has spring-loaded controls that can be recessed. (Ibanez; Visual Sound)

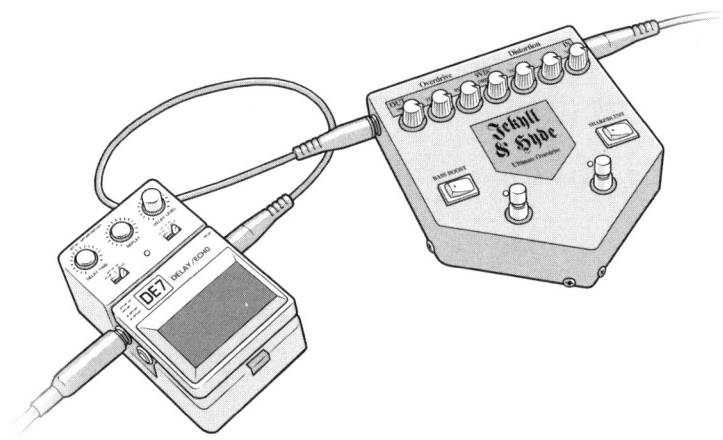

One or more
Most stompboxes house a single effect, but some offer effect combinations, such as auto-wah and distortion. If so, check if you can change their order, using auto-wah either before or after the distortion.

Modeling
Effect modeling is used in stompboxes too: A single distortion pedal may provide you with the models of a number of classic distortion pedals, phasers or flangers.

In, out and batteries
The most basic stompboxes have nothing but an input and an output, and a special input for an adapter (see page 210). Stompboxes can be battery powered too. Battery life depends on the type

202

of effect. A battery in a wah-pedal can last for three months, while a delay/reverb pedal might kill one in a couple of hours — a lot of money, considering the price of popular 9-volt batteries.

Which battery

Alkaline batteries are generally preferred as they last longer than regular ones. Some people use standard rechargeable batteries. These types of batteries help save both money and the environment, but they typically supply less power. There are more expensive rechargeable batteries that do work well in effects and related units.

Changing batteries

Opening the battery compartment may be as simple as pushing a release button with your finger or as time-consuming as unscrewing as many as four screws. Tip: Always unplug effect pedals when you're done. As long as a cable is plugged into the input, the battery is draining. Tip two: Some old fuzz and distortion pedals tend to sound best (warmer or dirtier, that is) with weaker batteries. You can emulate this effect by setting an adapter at a lower voltage than indicated (but this may damage the unit and void your warranty). Some pedal boards have variable DC jacks that allow lower voltages for this purpose too.

Pedal board

If you use a number of daisy-chained effect pedals, you may consider getting a *pedal board* that holds pedals, tuners, and other devices in place, preferably using a central power supply or *multi-adapter*. Ready-made pedal boards are available, but you can make one yourself too. The most basic version is a plank with Velcro patches to attach your units to. Ready-made designs may include a lid, surge (power spike) and short-circuit protection, noise filtering, power conditioning, storage space for cables and accessories, and other features. They also weigh less and look neater than most homemade boards.

Expression pedal

A volume pedal is the most basic example of a real-time *continuous controller*. Such pedals are also known as *expression pedals*. A wah-wah uses the same type of pedal, which is why there are so many combined wah-wah/volume pedals. Depressing the front end of the pedal switches from one effect to the other. Check how far the pedal can move up and down: The larger this distance or *throw*, the finer the control it will offer. Multi-effect units often have one or even two expression pedals on board (see page 207). Models that don't may have a special input for an external pedal.

Expression pedal.

MULTI-EFFECTS UNITS OR STOMPBOXES?

Pricewise, multi-effects units seem much more attractive than stompboxes: You can buy a good unit with dozens of effects for the price of two or three dedicated effect pedals. On top of that, a multi-effects unit is programmable (if you want to use a stompbox for different sounds, you'll need to memorize or jot down the various settings). So why do people buy stompboxes?

Why stompboxes?

- They're easier to use, featuring a separate control for each parameter. **No menus** to go through!

- Their higher price often buys you a **higher quality** effects.

- They typically provide **better control** over the parameters.

- They allow you to combine your **favorite chorus** with your favorite distortion.

- Stompboxes generally **last longer**.

- You can set them up **in any order** you like.

- You're not stuck with the **overall sound character** of a multi-effects unit.

But...

That said, it's hard to organize a multi-pedal setup as well as multi-effects unit. Also:

- Multi-effects units also allow you to experiment with **a large variety of effects** for a very reasonable price.

- They allow you to change the settings of a number of effects with a **single footswitch**, so there's no need for tap dancing.

- There's **less hassle** with cables, adapters, and connectors.

- They often provide you with **lots of extras** such as a built-in tuner, connectors, and loads of other features, as listed earlier in this chapter.

What you need

Rather than making a choice based upon the lists above, you can also simply look at what it is you need. If you use different effects and sounds in every song you play, a programmable unit will be your best choice. But if all you ever need is a chorus and a delay, you're probably better off with dedicated pedals.

Parameters

A larger number of parameters allows for more control over the sound. Note that basic multi-effects units typically offer less

205

parameters than stompboxes do, while advanced units may provide more options than traditional compact effects.

MULTI-EFFECTS UNITS

Multi-effects units differ in pricing, features, overall sound quality, and the quality of the various effects (all effects can be great, or some can be good and others less so), in their construction and user-friendliness, and more. Here are some things to consider:

Number of effects
A large number of effects (with several reverb types, chorus types, and so on) doesn't necessarily make you happy, but it can increase your chances of finding the one you're looking for. Less may be more, however: At the same price, a unit with less effects and features is likely to offer better sound and overall quality.

Combining effects
Some units offer preset effects combinations only (e.g., chorus+ compression, phasing+delay, etc.). Being able to choose your own combinations, as well as the order of the effects in the chain, makes a unit a lot more flexible. The number of effects you can use simultaneously varies from two to ten or more.

Presets
The number of presets varies from a few dozen to a hundred or more.

- Some units have separate sets of editable user presets and **noneditable factory presets**, so these 'example settings' never get lost.

- With other units, all presets are preprogrammed, and you can **edit all of them**. You may be able to restore the original settings either one by one, or globally. In the latter case, you will of course lose all changes you made to the factory presets.

- A few models have no **factory presets** at all, leaving things

entirely up to your own creativity — implying that you should really be creative.

- Factory presets, which are often created **with the help of well-known musicians**, are mainly designed to show what a device is capable of. You will usually have to adjust them to match your instrument, your amp, your band, and the music you play.

Controls

A large number of controls can make a unit look daunting, but such units are usually easier and quicker to operate than the ones that just have a few multi-function buttons. On the latter, you may have to push a button seven times to get where you want — going through menu after menu as you do — while dedicated buttons allow you to change settings right away. This is most important in live situations.

TIP

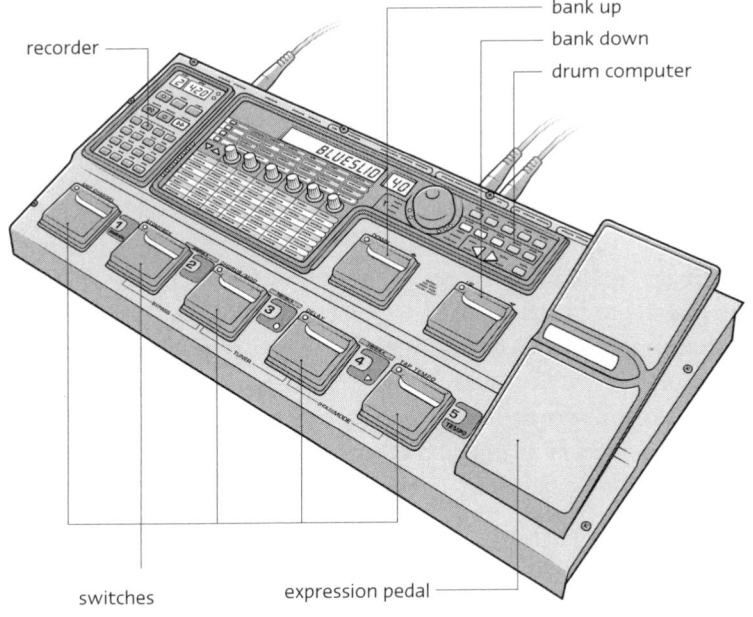

recorder

bank up
bank down
drum computer

switches

expression pedal

*Floor-based multi-effects unit with expression pedal, recorder, MP3 player, drum machine and much more.
(Digitech)*

207

Floor-based multi-effects processors

Live use requires presets that are easily accessible. They're usually stored in a number of banks, each bank holding a number of presets. Larger floor-based units often have two (one up, one down) pedal switches to scroll from bank to bank, and a series of pedal switches to directly select the presets from the active bank: If each bank contains five presets, you would have five of these switches.

Expression pedal

Floor-based units often have one or even two integrated expression pedals, or they have an input for an optional expression pedal. The pedal can commonly be used to control a number of assignable effect parameters: You can use it as a volume or a wah pedal, or to control your tremolo speed or distortion level, for example. Some allow you to gradually switch from one preset to the other. *Tip:* You may be able to set minimum and maximum values or *stop points* for each parameter, e.g., a minimum and a maximum volume level.

(MODELING) PREAMPS

If you're not satisfied with your guitar amp's lead sound, you can get a distortion pedal, or you can get a floor-mounted tube-driven preamp for the same purpose. These designs usually offer two or more channels as well, almost turning your old amp into a new one. Examples are shown on pages 102 and 183. Preamps can be hooked up directly to a mixing board, so you can always bring your own sound to a gig or a recording session, in a very portable format.

Modeling preamps

Next to a number of amp and cabinet models, most modeling preamps also house a variety of digital effects.

Versatile

Modeling preamps can be used live, but they're great practicing and recording tools too, supplying you with the sound of several

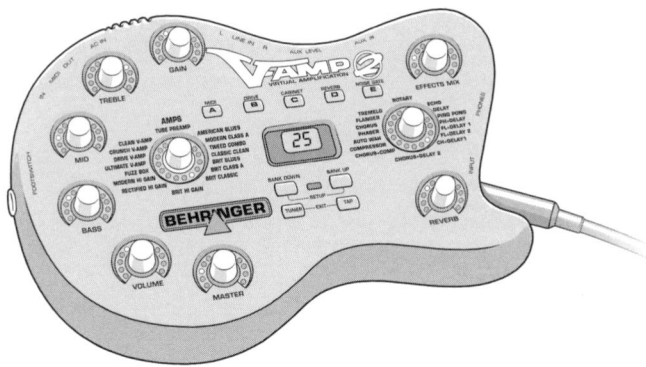

A modeling preamp.

amps and cabinets at a very affordable price. They're available for guitarists and bassists, the first type also being used by keyboardists and other musicians.

(Guitar) amp?

- Modeling preamps have built-in **speaker simulation** (see pages 56–57), so you can plug them straight into a mixer or recording console.

- Likewise, you can use the modeling preamp with a **keyboard amp** or a set of powered speakers.

- If you use a guitar amp, you'll usually get the best results if you feed the signal to the amp's **return input**. Turn the speaker simulation off; you're using a real guitar speaker now!

- Modeling preamps usually have **separate outputs or settings** to connect the unit to a guitar combo amp, a stack, a keyboard amp, an effects loop, a line-level input (PA, recording console), or a pair of headphones.

RACK EFFECTS

Rack effects units can house just one effect, or any number of

209

effects and effect combinations. The smaller the number of effects, the more control you usually have over them, and the higher their quality often is. If you want to use a rack effect live, see if it allows for pedal operation.

ADAPTERS

Stompboxes use batteries or an adapter; most multi-effects require an adapter. Adapters typically convert the AC current from the power outlet to DC current that your equipment uses, and it lowers the voltage. Very few effects units come with a built-in power supply, which is more expensive, but less vulnerable and more practical. *Tip:* Adapters are also known as power supply units (PSU).

Dedicated adapters
Some companies force you to use their dedicated adapters by using special plugs; others require you do the same, stating that the use of another adapter will void your warranty. That said, universal adapters may work just as well, as long as they're set to the proper voltage and polarity. Preferably use a stabilized adapter. Some companies make special adapters for the effects of other brands, with or without noise- or hum-suppressing features.

> ## Polarity
> *Most devices require center-negative polarity, meaning that the center of the adapter's plug is negative (cold, -), and the barrel is positive (hot, +). Reversing the polarity may damage the device, or it just won't work. An adapter's polarity may be user-switchable.*
>
>
>
> This icon tells you that center-negative polarity is required.

Phantom power

In 2009, P3 (Phantom Powered Pedals) introduced a kit that allows you to remotely power your pedals, similar to the way condenser mics are phantom powered.

MISCELANEOUS FEATURES

Here are some more things to pay attention to when checking out effect equipment:

Inputs and outputs

Most of the inputs and outputs on effects units are not different from the ones on amps, which are dealt with in chapters 2 and 4. Some units have separate outputs for the effected signal and the uneffected, dry signal, so you can send a dry signal to the mixing or recording console (where it can be processed for the room or the recording), while you listen to the effected signal — adjusted to your own taste.

Digital outputs

Units that are designed for studio use may have *digital* (AES/EBU or S/PDIF) *outputs* that allows for noise-free signal transmission.

MIDI and USB

Multi-effects and modeling amps often have MIDI ports, and a growing number have a USB or Firewire port to connect the unit directly to your computer. USB type ports also tend to replace the traditional five pin MIDI ports, enabling faster data transmission.

Display

Programmable units have a display. The larger it is, the easier it is to read, and the less you'll have to deal with incomprehensible abbreviations or codes, rather than full names. On many units, you can name your presets. If a unit has an LCD display (like a cell phone), check if it's backlit so you can still read it on dimly lit stages.

211

And more...

- Again, some of the tips in **Chapter 6** apply to effects units too.

- **Non-skid rubber** pads or a rubber base keep floor-mounted units where you put them.

- If an effect is switched off, it **shouldn't affect the signal** at all. Some units offer a *true bypass* that makes the signal pass by untreated.

- A **status LED** on a stompbox shows whether the unit is on or off. It usually also warns you when battery power is running low.

- Check if the unit has a **strain-relief clip** that prevents the adapter from getting unplugged inadvertently.

- Multi-effects units and modeling preamps usually have a **built-in tuner**. Their quality differs. A tuner with a display that uses just three LEDs (flat, in tune, sharp) doesn't make for easy tuning.

- There are **tuners in a stompbox housing**, designed to be used between your instrument and your first effect or your amp. A mute switch allows you to tune silently.

- See how well the controls on floor-based multi-effects are protected against misplaced feet. For example, some units have a **bumper bar** between the pedal switches and the other controls.

- A **phrase trainer** is a built-in or external device that can record a lick or a section of a song and play it back slowly, yet at the original pitch, allowing you to figure out what's being played.

- For complex multi-effects units, ask to see **the manual**. This might help you to get an impression of the unit's features and user-friendliness (and of the quality of the manual as well).

- An infrared transmitter lets you use a pair of **wireless** infrared headphones.

- Some units have a bass or center **canceling feature** that filters out the bass, guitar, or vocal part (to some extent, that is), allowing you to join in with the original recording.

212

- Taking things on the road? A **padded pouch or a carrying case** may be included with the unit. If not, you can buy these useful accessories separately.

- If you buy a second-hand effect pedal, check the battery compartment for traces of **battery fluid leakage** that may have damaged the unit.

- Guitar synthesizers and other units may require a special type of **hexaphonic or hex pickup** that can be retrofitted to your bass or guitar.

14

Microphones and Pickups

If you're a singer or an acoustic musician, an amp
isn't complete without a microphone or a pickup. The
information in this chapter makes selecting the right one
a lot easier.

A microphone is somewhat like a speaker in reverse. Under the microphone's grille is a *cartridge* or *capsule* that converts vibrating air — the sound from your voice or your instrument — into electric energy that can be sent to an amplifier. The two most popular types of microphones for onstage use are dynamic and condenser microphones.

Dynamic

A *dynamic microphone* has a dynamic cartridge with a thin diaphragm that responds to sound by making a coil move in a magnetic field. This creates electric energy that can be amplified.

Condenser

In a *condenser* microphone, the cartridge has two extremely thin, electrically charged plates: the *diaphragm* and the *backplate*. Sound makes the diaphragm or *membrane* move, creating variations in the spacing between the plates, which generates electric energy.

The difference

Dynamic and condenser microphones may look very much alike, but their cartridges make them behave differently. Compared to the diaphragm in a condenser microphone, the moving coil in a dynamic microphone is quite bulky: The condenser's diaphragm responds much faster, which translates to a more precise, direct, crisp sound.

Sensitive

A condenser is also more sensitive, providing enhanced detail and transparency. The response of a condenser mic tends to be more uniform too, bringing out the natural timbre of the voice or the instrument, rather than coloring it. Studios often use special large-diaphragm condenser microphones.

Rugged

Dynamic microphones are known to be very rugged and reliable, making them ideal for life on the road. They produce a warmer, fuller, or smoother sound, which is what most rock and jazz singers are after. Dynamic mics can often handle high sound pressure levels, so they're commonly used for drums and amps too.

Phantom power

There's one more technical difference between condenser and dynamic models: Condensers need power. This is usually supplied by the mixer's microphone preamps (phantom power; page 164). The small instrument-mounted mics that are covered later in this chapter commonly take their power from a battery in an external preamp.

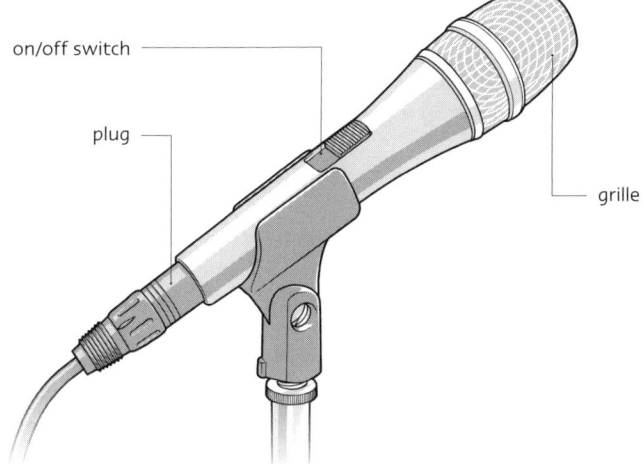

A vocal microphone. (Sennheiser)

on/off switch

plug

grille

Microphone preamp

If you plan to buy a condenser microphone, it may be wise to also invest in a microphone preamp. This allows you to also use your mic if the available equipment does not feature a microphone input and/or phantom power: you can simply connect your preamp to the line input (instrument input) of a keyboard amplifier, for

TIP

Ribbon microphones

There are more types of microphones than the ones described above. One example is the ribbon microphone, which uses a thin metal ribbon in a magnetic field. Ribbon microphones usually have a warm and natural sound. Because they're so vulnerable, they're typical studio microphones.

217

example. Microphone preamps with phantom power start at less than a hundred dollars. They often feature a limiter as well. Studio quality microphone preamps may cost thousands of dollars.

PICKUP PATTERNS

A mic's *pickup pattern* indicates how it responds to sounds coming from different directions. Pickup patterns are also known as *sensitivity fields*, *polar* or *response patterns*, or *directivity responses*.

Uni, omni, and bi

- Most microphones have a **unidirectional pattern**: They respond to sounds from one (*uni*) direction only. This means they can be focused on one instrument or voice, without other sources bleeding into them.

- **Omnidirectional microphones** pick up sounds from all directions, which makes them very likely to cause feedback. As a result, they're rarely used onstage.

- **Birectional microphones** accept sound from two directions, front and back (e.g., ribbon microphones).

Patterns

Unidirectional microphones come with different patterns.

- Many microphones have a **heart-shaped** or **cardioid pickup** pattern. As the illustration on page 219 shows, this rejects sounds that come from the rear, where a single monitor would be. Off-axis sound is reduced.

- The next popular design is the **supercardioid** pattern, with a tighter heart shape that is focused on the front (at 0°), with a bit of sensitivity on the rear (at 180°). It's least sensitive around 120°. This is very effective on stages with a PA system.

- A **hypercardioid** pattern has an even tighter pickup angle. *Tip:* Tighter pickup angles require better microphone technique, as

even smaller variations in the mic's position will influence the
sound.

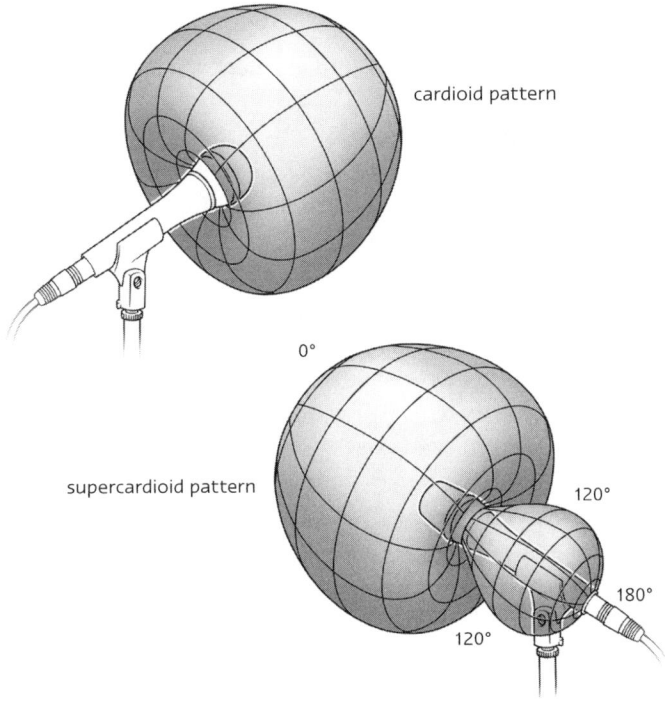

cardioid pattern

*Cardioid and
supercardioid
microphones
reject and
reduce
sound from
various angles.*

0°

supercardioid pattern

120°

180°

120°

Switchable

Some mics, mostly studio condensers, have two or more polar
patterns that can be selected with a switch, or you can change the
pattern by mounting a special cap on the mic.

Feedback rejection

The narrower a microphone's pattern, the less likely it is to cause
feedback. That said, even microphones with similar patterns will
differ in how well they reject this ear-damaging effect. If you
play often in feedback-sensitive environments (loud band, lots
of speakers, small rooms, hard walls), search for microphones
that claim to have 'high gain before feedback' or 'great feedback
rejection.'

219

AND MORE

Dealing with microphones also requires you to deal with frequency responses, sound pressure levels, impedances, and other data.

Frequency response

Like speakers, microphones have different frequency responses. This makes them sound different from one another, accentuating or suppressing different frequencies. Singers often prefer a specific microphone to either compensate for or emphasize certain characteristics of their voices: Mics can make you sound warmer and fuller, or brighter and more aggressive. *Tip:* Some microphones offer a variety of selectable frequency responses. A *linear* or *flat frequency response* means that a mic doesn't color the sound at all.

Sound pressure level

A microphone's sound pressure level (SPL) indicates the maximum sound pressure it can handle. These SPL values, expressed in decibels, are a lot higher than the decibel figures listed on page 82, because microphones are often very close to the source. Ever had someone shout at you at the same distance that most singers hold their microphones? Then you know why an SPL of 150dB is not unusual for a vocal mic. Drum mics can go up to 160dB. Some microphones have a switch-activated pad that attenuates the signal by (usually) 10dB, allowing them to handle even louder sounds.

On/off

An on/off switch can be very practical, but the microphone may be off when you hope it's on, or vice versa (which can be even more painful). On some mics, the switch can be locked, and it's often designed so that you can't turn it on or off inadvertently. Check if the switch is noiseless: You shouldn't hear it through the speakers.

Unbalanced

Low budget mics often have a permanently-attached cable with a ¼" phone plug. These are high-impedance, unbalanced mics that are prone to picking up noise.

Professional

Professional microphones have a detachable cable with XLR connectors. Being low-impedance and balanced, they're much less sensitive to noise and they allow for long cable runs.

Proximity

The closer you get to a microphone, the bassier the sound gets. You can use this *proximity effect* to your advantage, adding warmth to the sound — but you can also reduce it with a *low-cut* or *bass roll-off switch* that cuts unwanted rumble too. The proximity effect differs with each microphone.

Accessories

Some microphones come with a matching clip or holder. There are different holders for straight- and tapered-shaft mics. The holder's thread may be different from the mic stand's thread (there are two common threads, ⅜" and ⅝"). A *thread adapter* solves the problem.

MULTI-PURPOSE AND DEDICATED MICS

Microphone makers always specify the applications their microphones have been designed for. Some designs are extremely versatile (one mic that can handle dynamic vocals, subtle woodwinds, powerful brasswinds, screaming guitar speakers, or loud snare drums equally well); others were made with a specific application in mind. Online and hard-copy microphone catalogs and can help you narrow down your selection.

Vocalists

There are many dedicated vocal microphones. Please check out the microphone chapter in *Tipbook Vocals* (see page 289) for additional

221

information, including tips on pop filters, microphone technique, headsets, and more.

Instrument microphones

Many horn players use vocal microphones. This isn't always the best thing to do. For example, a unidirectional (vocal) microphone may not register the full sound of a saxophone or a clarinet, which radiates not only from the instrument's bell but from all its open toneholes as well. Trumpets, on the other hand, easily produce a sound pressure level that can be too much for a typical vocal microphone. Dedicated instrument microphones are better suited for these instruments.

Special mics

Also, there are dedicated microphones for other applications, from miking guitar amps to blues harps (the latter with a very pronounced, telephone-like frequency response). Special bass drum mics are designed to deal with high sound pressure levels. Similar microphones are used for bass amps and tubas, for example.

Dynamic range

A regular stand-mounted microphone allows horn players to increase their instruments' dynamic range by moving to and

A clarinet with two dedicated clip-on microphones and a belt-clip preamp. (SD Systems)

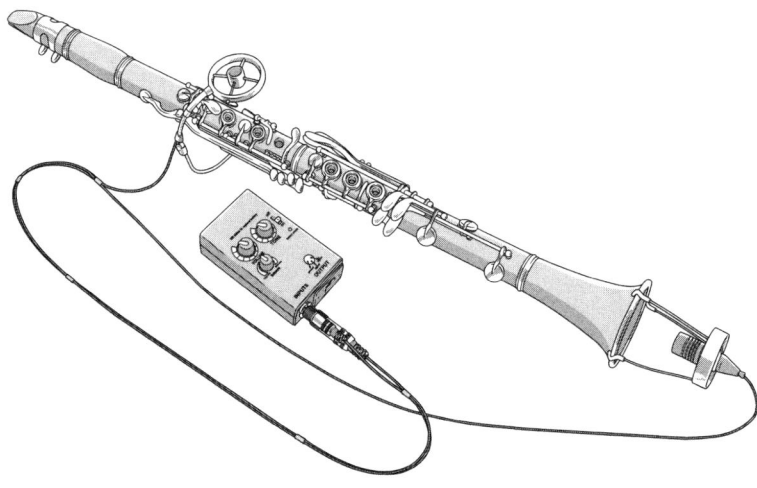

away from the microphone. Conversely, using a stand-mounted microphone limits their freedom of movement.

Miniature microphones

If, as a horn player, you want to be able to move around, you may consider getting an instrument-mounted microphone. These are usually small condenser mics, mounted on adjustable clamps or goosenecks that allow you to adjust their position. The mic's preamp and power supply are housed in a small belt-clip box that commonly has tone and volume controls. Systems that use two mics usually have a balance control too. Miniature mic systems (both wired and wireless) are available for saxophones, flutes, trumpets, and other woodwinds and brasswinds.

Bowed instruments

Similar systems have been designed for bowed instruments. To help prevent feedback, they're often combined with a pickup system (see pages 225–227). Pickup/microphone combinations for acoustic guitars are available too.

TESTING MICROPHONES: TIPS

Buying a microphone is mostly a matter of using your ears, trying to find the one that makes you or your instrument sound the best. Unfortunately, it's very hard to test microphones for drums or guitar amps and other applications in a music store, so you'll have to rely on reviews, experiences of other musicians, and the salesperson's advice. If want to try them out live, see if you can borrow or rent microphones.

Prices

While the pricing of most classic, major-brand microphones is pretty stable, various Asian companies have introduced very decent mics for very attractive prices.

• You can get a complete five-piece microphone set for **drums** for less than two hundred dollars, for example.

223

- Note that **top-of-the line microphones** do not need to be that expensive, however, and that good mics can last for many years, if properly cared for.

- Popular, professional **vocal microphones** can be purchased for less than a hundred fifty dollars — and they're used by the stars!

- A good **vocal condenser mic** may cost you four hundred dollars or more.

- **Miniature mic systems** typically range from one hundred fifty to more than six hundred dollars.

> ### Studio mics
>
> High-end studio mics (especially tube mics) easily cost thousands of dollars. That said, there are pro models for less than a thousand dollars, and large-diaphragm mic for home studios are available for less than two hundred dollars.

More tips

- If you have to audition mics in a store, see if it's possible to bring a **recording of your band** that you can play or sing along with, so you can hear what they sound like in their proper context.

- When trying mics out, either live or in a store, it's best to hook them all up to **a mixer**, and adjust them so they're all equally loud.

- Set all **EQ controls flat** and don't use any effects to begin with.

- Many of the testing tips in **Chapter 6** can be applied to microphones too!

- If you already have a microphone and you want to replace it with a better one, bring it along: **Your own microphone** is a great reference point.

- Sing and play the way you normally do: **high and low notes, loud and soft**. Your loudest notes shouldn't distort; the softest words should be picked up impeccably and directly. If you get gradually louder, the amplified sound should do so too. Low notes should sound clear and not too bassy, and high notes should have body.

• Microphone brochures and online catalogs often contain more technical data than you can handle. Manufacturers probably put all those numbers in because it's easier to indicate sound in figures and frequency curves than it is to describe it in words. Judging a microphone by its technical data requires a lot of knowledge and experience, so you're probably better off **trusting your ears**.

• For home recording, dedicated **USB microphones** are available as well.

PICKUPS

Amplified acoustic guitars, cellos, and other string instruments usually sound best when a microphone is used. However, the instruments' bodies make them extremely sensitive to feedback, and the mics can pick up lots of background noise. Using a *pickup* can solve those problems.

Acoustic/electric guitars

Most acoustic/electric guitars use a pickup in the form of an ultrathin strip of pressure-sensitive material that is mounted under the guitar's saddle. The control panel for the built-in preamp is usually located on the shoulder of the instrument. Some guitars have a miniature preamp that's built into the guitar's output jack. Pickups and preamps can be retrofitted to any guitar. There's more on guitar pickups in *Tipbook Acoustic Guitar* (see page 285).

Other instruments

There are special pickups for all kinds of instruments, from pianos and violins to drums and ethnic instruments, in different shapes and formats, and using different materials and techniques.

225

They're often attached to the instrument's soundboard with a self-adhesive backing (*soundboard transducers*). For some instruments, so-called strip microphones are available: Ultrathin, phantom-powered condenser pickups that behave very much like a condenser microphone. Even the best pickup will only perform well if it has been properly installed; professional assistance and/or installation is required for some types of pickups, and recommended for others.

Violin, cello, bass

The most basic system for bowed instruments is a single pickup or *sensor* that is mounted between the wings of the instrument's bridge, or between a bridge foot and the body. For a more uniform overall string response, many systems use two pickups, and some use four: one for each string. Other designs have the pickup built into the bridge, which then needs to be adapted to the instrument by a luthier.

Violin with a pickup in the bridge (Fishman)

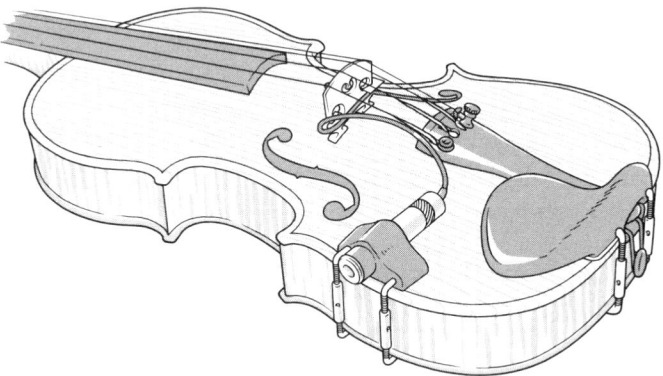

Pizzicato and arco

Upright bass players often find that pickup systems produce a fine bowed (arco) sound, but leave something to be desired when plucking the strings (pizzicato), or vice versa. To solve the problem, there are systems that have independently adjustable presets for both playing styles, usually using two or more pickups.

226

Variations

Pickup systems vary in price and sound quality, and in how easily the pickup can be removed and reattached (an important issue for many string players). Some systems make the instrument sound very detailed, as if you were really close, while others produce a more distant sound with more body. Another difference is that some are more sensitive to feedback than others, the latter being better in high-volume situations. Like a good microphone, a pickup system should be properly shielded to prevent hum and radio interference.

Pickup and microphone

Piezo pickups are often said to lack warmth, or even to sound quacky. For a more natural, classic sound, the piezo sensor is often combined with a small microphone. A small preamp/mixer allows you to balance the signals from both sources: more mic in low-volume situations, and more pickup if it needs to get louder. The microphone can be attached to the tailpiece on violins, or to the bridge on upright basses.

Magnetic pickups

Acoustic guitarists may also use a magnetic soundhole pickup. Magnetic pickups respond to the movement of steel strings, just like the pickups on electric guitars. Some soundhole pickups come with an integrated condenser mic that adds the natural resonance of the instrument's soundbox to the sound. Magnetic pickups for other instruments with steel strings (e.g., banjos) are also available.

Prices

Prices vary from around a hundred dollars for a basic violin pickup system to four hundred dollars or more for pickup/microphone combinations.

Preamps

For many types of pickups it's advisable to use a preamp that matches the pickup's impedance to that of the amplifier input. Please see pages 88 and 133–134 for additional information.

15

Cables and Wireless Systems

A short chapter on the main types of cables, what makes a good cable, and the main features of wireless microphone and guitar and bass systems.

Apart from all sorts of special cables, there are three main signal cable types:

- **Instrument cables** (a.k.a. guitar cables) connect instruments, amps, and effects.

- Balanced **microphone cables** connect mics to amps or mixers. The same type of cable is also used for other balanced connections.

- **Speaker cables** connect amps to speakers.

Identify
You can usually identify instrument and microphone cables by their plugs: They have ¼" phone plugs and XLR plugs respectively, though some mic cables have one of each. Speaker cables may confuse you, as they come with both types of plugs. Only when a speaker cable has Speakon plugs (see page 55), is it clearly identifiable as such. To help you out, cable makers often print the cable type on the cable's jacket.

The right cable
Always use the right type of cable for the job.

- Using speaker cable with ¼" phone plugs for instrument connections causes **excess noise**. Speaker cables are not shielded (see page 232).

- Instrument and mic cables are **way too thin** to connect speakers with. You'll lose a lot of power and you may even burn the cable or damage the amp or the speakers.

- A speaker cable with XLR plugs will **only produce noise** when using it for a microphone.

Can you hear the price?
A high-end cable can be five to ten times as expensive as a cheap cord of the same length. Do they really sound different? There's only one way to find out: Compare them and listen, preferably in a blind test, using your own equipment. With instrument and speaker cables shorter than ten feet, it's unlikely you'll be able to tell the difference.

230

Spend more

Even if every cable sounds the same to you, it may be worthwhile to spend a little or a lot more, with better cables offering more rugged construction, better plugs, better strain relief, lasting flexibility, less tangling, less handling noise and interference, impact-absorbing insulators, etc. For home practicing, a low-budget cable will usually do.

Tip: Always get the shortest cable that suits your needs in order to avoid noise and/or signal loss. With balanced (microphone) cables, length is not an issue.

Jackets

A cable's outer jacket is usually PVC, a flexible material that has no 'memory.' Some prefer a woven or braided outer jacket for improved durability. Both materials are available in a variety of flashy and vintage designs and colors, as well as black, of course.

Stranding

All cables have stranded wire conductors. Thinner strands do not influence sound quality, but they do increase the cable's flexibility and longevity.

Plugs

The main types of plugs have been dealt with in Chapter 4. The quality of a cable always depends on the quality of its plugs too. Many good cables use plugs made by companies such as Neutrik and Switchcraft.

INSTRUMENT CABLES

Instrument cables usually come in lengths from about six inches to 30 feet, the shortest ones being used for daisy-chaining effects. Right-angle plugs can help save space between effects units, and they fit instruments with a flush mount output jack better than a regular straight phone plug.

Too long

Long instrument cable runs make an instrument sound duller or muddier, less dynamic and less defined. The higher the *capacitance* of a cable, the stronger these effects will be. Better cables usually have a lower capacitance (some 80pF/m or less, for those who like figures). *Tip:* Many guitarists happen to like long cables for the way they degrade the signal. The late Jimi Hendrix was one of them. Another tip: The signal of keyboards, synths and other low-impedance sources will not be bothered by longer cable runs — but longer cables may be noisy.

Shielding

Instrument cables transport weak, unbalanced signals that are sensitive to various types of interference (RFI and EMI, for example), as well as to microphonics. Various types of shields can be used to fight these effects. The conductor is the hot lead (+) in the cable; the braided copper-wire shield that surrounds it doubles as the cold lead (–).

Instrument cables are coaxial cables.

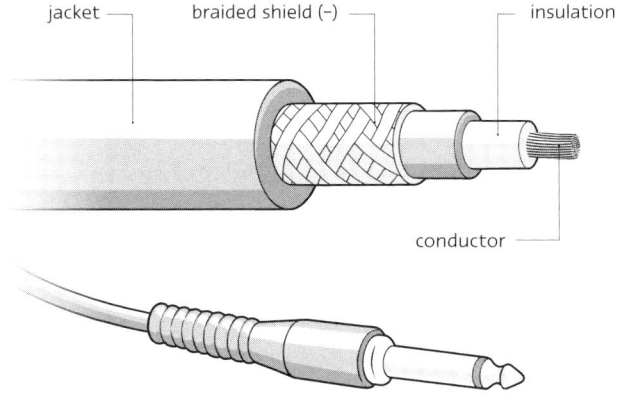

jacket — braided shield (–) — insulation

conductor

Noise

The longer the cable you use, the more noise it will pick up. If the total cable length you use (instrument to effects, effects to amp) exceeds 18 feet, you may consider a wireless system, or use a DI box that turns the unbalanced instrument signal into a noise-suppressing lo-Z, balanced signal — provided that your amp can handle that type of signal.

232

Handling noise

Handling a cable or stepping on it may create a cracking or slapping sound. Conductive PVC layers or other materials can help to reduce this *triboelectric impact noise*.

TIP

> ## Directional arrows
>
> Instrument cables sometimes have directional arrows, indicating the direction of the signal flow. Hooking them up the wrong way around will not make the sound suffer, but shielding may be less effective. No arrows? Some players always connect their cables so that the signal flows along with the direction of the print on the jacket. The least you can do is try this once to see if it makes a difference for you.

Bassists

There are special instrument cables for bassists, claiming an improved low-frequency response and a larger dynamic range. Trying them is the least you can do if you're serious.

Not really

Here are some things that are often said to be important. But are they?

- **Resistance** is not an issue when it comes to instrument cables; it *is* for speaker cables.

- The same goes for the cable's **gauge**. Speakers require heavy-gauge cables; instruments don't.

- Some claim that **oxygen-free copper** wire reduces internal corrosion and signal distortion, while others say these qualities have no effect on instrument and microphone cables. Back to reality: Most companies use this type of wire — because it's said to be easier to process — so there's not too much to discuss, really.

- **Gold-plating** helps prevent corrosion, but it will soon wear off on frequently disconnected plugs. A regular nickel-plated plug is fine.

233

SPEAKER CABLE

Speaker cables are very different from instrument cables, as they transport much stronger, very low-impedance signals. This makes them insensitive to interference, so there's no need for shielding.

Speaker cable.

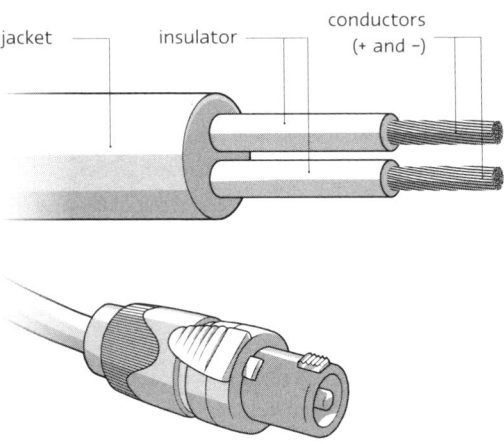

Gauge
The strong speaker signal benefits from heavy-gauge speaker cables, as these offer less resistance to this type of signal. Thin cables cause a loss of signal and reduce the speaker's damping, losing tightness and definition.

How much?
You'll be usually advised to use cable of 18 gauge (1 mm^2) or heavier. Note that a lower number indicates a thicker cable. For speaker cables longer than fifty feet, go to 14 gauge (2 mm^2) or

TIP

Wiring inside
No matter how short they are, the speaker cables that are used inside your amp or speaker cabinet influence your sound too. Having the original thin cables replaced by heavier-gauge ones can dramatically improve your sound.

234

heavier. *Tip:* Keep speaker cables as short as possible. Doubling the distance doubles your signal loss!

Powered speakers

Powered speakers use line-level signals rather than the much stronger speaker signals, so you use instrument cables to connect them.

MICROPHONE CABLE

Low-budget high-impedance mics (see page 221) usually have a permanently attached cable or a cable with an XLR plug on one side and a phone plug on the other. All other mics use a dedicated low-impedance, balanced microphone cable. Contrary to instrument cables and speaker cables, length is not an issue for these cables: A 200 foot microphone cable can sound just as good as a 20 foot one.

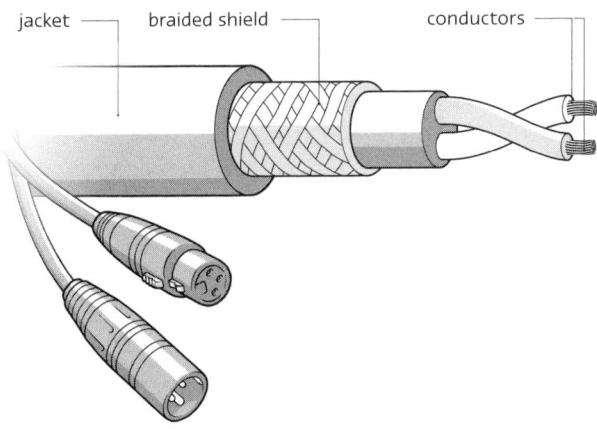

jacket — braided shield — conductors

A balanced (microphone) cable. One of the two conductors suppresses the noise that was picked up by the other, as explained on page 53.

Gauge

The type of signal in a microphone cable doesn't require heavy-gauge wiring. Should the cable be subjected to a lot of microphone-swirling, pulling, and twisting, a slightly heavier-gauge cable (think 20 gauge) may be considered.

235

Handling noise

Some microphone cables are more sensitive to handling noise than others. Always connect a microphone when you check cables out. Without it, any mic cable will suffer heavily from triboelectric noise.

WIRELESS

A wireless system allows you to move around onstage and frees you from long, tangling cables. An additional advantage for bassists and guitarists is that it prevents you from getting electrical shocks from your strings. See pages 242–243; Check!).

More or less money

Prices of wireless systems have come down, with starter sets for singers and (bass) guitarists being available for less than a hundred fifty dollars. Professional systems are much more expensive, buying you better sound and transmission, a larger range, improved noise and dropout reduction, less interference, rugged casings, etc.

The same?

With a professional wireless system, you shouldn't sound much different — if at all — than when you use a decent-quality cable. Affordable systems may reduce sound quality to a degree, and produce more noise. To find out what they do to your sound, just compare them to the cable you usually use, and remember that subtle differences will often get lost in the band's overall sound. Because they use radio signals, wireless systems are obviously more sensitive to interference than a regular cable connection, but there are different technologies to fight this, as you will see below.

Transmitter and receiver

Wireless systems consist of a *transmitter* and a *receiver*. Your mic or instrument plugs into the transmitter; the receiver is connected to the input you would normally use (amp, PA, etc.).

Transmitter

Transmitters have become smaller and smaller. The larger types are about the size of a cigarette pack. You plug your microphone or instrument in and attach the battery-powered transmitter to your belt or guitar strap.

Small

Smaller transmitters can be built into a microphone, and the smallest guitar-type transmitters are hardly bigger than a regular ¼" phone plug. These are powered by a AAA battery, and feature an on/off switch (which should be noiseless) and a status LED, as well as an input level control and a peak LED. A battery indicator, warning you when power is low, is a welcome addition. It may be a single LED or a more precise multi-LED battery-life meter. There are various models to fit the various types of guitar jacks.

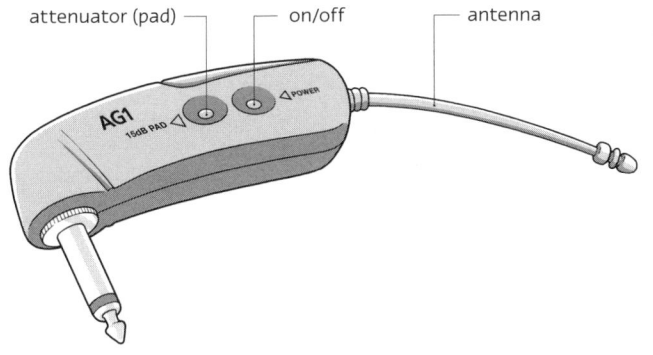

attenuator (pad) — on/off — antenna

A transmitter, hardly bigger than a regular plug. (Samson).

Battery life

Transmitter batteries may last anywhere from 12 to more than 36 hours. Please see page 203 for more information on batteries.

Receiver

Receivers come in various shapes, from rack-mountable devices to stompbox designs for guitarists. A multi-segment meter for monitoring audio level and incoming signal strength is a common feature. The more LEDs, the more precise the reading.

237

Receivers can be powered with an adapter and/or a battery. Both unbalanced ¼" and balanced XLR outputs may be present. The antennas, which receive the transmitter's radio signals, are usually of a retractable and folding design. Antennas are vulnerable, so definitely check their construction.

Bassists

Guitarists and bassists may use identical systems, but there are special wireless systems for bassists, promising a fatter bottom and a fuller sound than a guitar system. *Tip:* Most bass systems have a reduced response below 50Hz, while the low B on a multi-

A wireless microphone, a microphone with a plug-on transmitter, and a receiver.

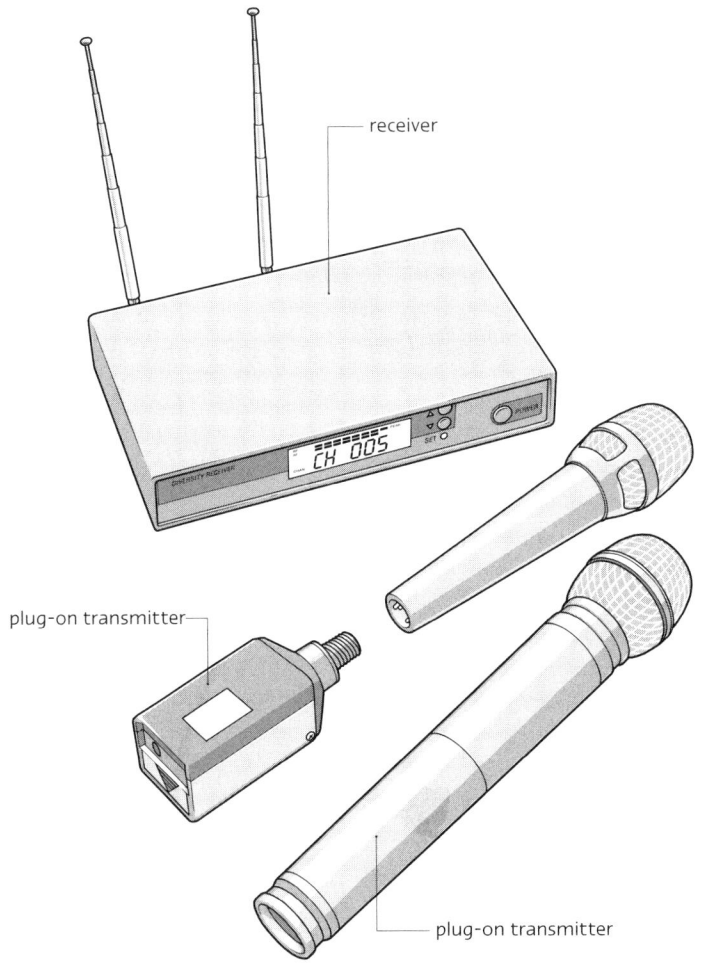

receiver

plug-on transmitter

plug-on transmitter

string bass vibrates at 30.9Hz. A problem? Not really, as the note's harmonics will still make you hear those lowest notes, especially in a band situation. Try it out!

Microphones

Many microphones are available in wired and wireless versions. Wireless versions are a bit bulkier due to the built-in transmitter and battery. If you have a wired microphone you like, you can turn it into a wireless model by getting a plug-on transmitter and a receiver. That way, you don't have to get used to another microphone, and it's often less expensive than a complete wireless system. Performance quality might decline slightly, however, or the sound may change — for better or worse. The transmitter also increases the mic's weight and size, altering its balance.

SOME JARGON

Most wireless systems use either the *VHF* or *UHF* frequency range for signal transmission. UHF systems typically cost more, but they're less sensitive to interference, power output is higher, the transmitter's batteries typically last longer, UHF systems have a larger operating range (although a VHF system allows for distances of two hundred feet or more), and they typically sound better.

Multiple frequencies or channels

A larger frequency or channel selection helps avoid interference with local broadcasting stations, and it allows for the use of multiple wireless systems without one bothering the other: It reduces the risk of *intermodulation distortion*, a result of interacting signals. Receivers may offer from 3 to more than 12 selectable frequencies.

Noise and interference

All kinds of filters and noise suppressors are used to suppress noise and interference. *Squelch* is a noise suppressor that kicks in when the signal drops below a certain (adjustable) level.

239

(Non-)diversity

Dropouts can be reduced or eliminated by using multiple antennas and channels, with the receiver automatically seeking out the strongest radio signal. A *diversity* system has a single receiver channel and two antennas; a *true-diversity* system uses both two antennas and two receiver channels. *Non-diversity* systems, with just a single channel and a single antenna, are most susceptible to dropouts. *Predictive diversity* means that a processor 'predicts' that the signal on one side will soon get too low, so it switches to the other side right away.

16

Care and Maintenance

Amplifiers, effect devices, microphones, and cables don't require much maintenance, but proper care will make them last longer and sometimes even perform better. The safety tips in this chapter are an extra: They do not replace the instructions in your equipment manuals.

Amps and effects units usually have the words 'Caution, risk of electric shock' and 'Do not open — no user serviceable parts inside' printed on their housing. Take these and similar messages seriously. You risk personal injury if you don't, and tampering with the equipment often voids your warranty.

Power off, volume down

Before connecting a unit, always switch all units off and turn their volume controls down. This prevents sudden bursts of sound as well as damage to your equipment and your ears. The amp should always be turned on last. Done playing? First switch off the amp, then the other units.

One power circuit

Make sure all amps, effects, and other devices use the same power circuit. This helps prevent hum. That circuit should not be shared with washing machines, refrigerators, fluorescent lights, air conditioners, and other noise-generating devices.

Grounding

Always use a grounded power outlet for equipment with a three-prong power plug. The third prong connects the unit to the ground, reducing the risk of electric shocks.

Check!

Your equipment will not be grounded if the power outlet is not

Warning

Unfortunately, grounding or earthing may generate ground hum. Never break or cut off the ground prong to solve this, and do not use a ground-lift plug. Basically, the only safe remedy is a ground lift switch (see page 61). If there's no such switch available, remove the earth connection of the XLR cable at one end, or have a technician do this for you. Such cables are available readily-made as well. Note that this solution may induce high frequency noise.

242

properly wired. A three-prong outlet tester helps you check this for just a few dollars. Too much of a good thing? Not if you know that guitarists have died because of lethal voltages from their strings. Also make it habit to periodically check your power cables and their plugs.

Circuit breaker

A circuit breaker is another safe and affordable investment. It cuts off the power if something goes wrong. Many devices (from amps to multi-outlet power strips) have one built in.

PREVENTION

There's more you can do to prevent problems. First of all, don't put glasses, bottles, and other liquid containers on your equipment. Should liquid get into a unit, switch it off immediately and have it checked. Do the same if a unit seems to malfunction. Always handle electrical equipment and power cables with dry hands!

Cool

Amps get warm, so don't place them near other sources of heat, and see to it that units with cooling sinks and fans have plenty of room to dissipate their heat. Don't put tube amps with open backs against a wall, and don't close the back to see what that sounds like: It won't sound that way for long.

Warming up and cooling down

If your equipment comes in from the cold, wait a while before using it. Tube amp? Wait half an hour before you play, and wait another half-hour before you take it out into the cold again. Ideally, tube amps should cool down before being moved anyway: Like light bulbs, hot tubes are very sensitive.

Speakers

Preserve your speaker cones: Don't store accessories in open-backed enclosures (see page 115) unless you can attach them.

243

Preserve your speakers: Don't drop them, try to avoid feedback, and use your volume control wisely. Rattle? Check the nuts that hold the speaker(s). When one or more are loose, don't overtighten them.

Cover

Electronic equipment is sensitive to moisture and dust. Both may attack the electronic circuitry and moving parts (e.g., pots and switches), and dust eventually blocks ventilation openings. See if these openings need vacuuming from time to time, and keep your equipment covered when not in use. Covers are available in various sizes, and some units come with their own.

TIP

Batteries

Unused batteries may start leaking fluids that can damage your equipment. To prevent this, remove batteries from units that you're not going to use for a several weeks or longer and don't forget to take out the batteries when you mount pedals on a pedal board.

Cables

Grab the plug — not the cable! — when connecting or disconnecting cables. The internal conductors and shielding get damaged over time when cables are pinched, sharply bent, twisted, or squished. Elbow-wrapping cables means twisting them, so don't. Instead, loosely coil them up and tie them. Some use tape for this purpose (PVC-tape is popular. Other types of tape may leave a residue, and all types of tape stick to the stage or your shoes), but most players prefer special Velcro-style cable ties.

Coiling up cables

There are numerous ways to coil up cables, and describing them mostly leads to utter confusion. One of the most basic ways is as follows: Find the middle of the cable, and make sure the cable is not twisted. Then hold the plugs in one hand, and fold the length of the cable a number of times. *Tip:* Let the plugs stick out or plug

244

XLR jacks into one another: This prevents the cable from tangling. Tie the cable into a loose, single knot when its folded length is about 30" to 40", or use a cable tie instead.

Cables onstage

Avoid stepping on your cables onstage, and place them so that people can't trip over them. Use stage tape or gaffer's tape to attach cables to the stage floor. White or silver tape may help keep others from stepping on them. Coil up excess cable. Don't hang the coil over a plug, as this may wear out the plug or the jack. Try to avoid using adapter plugs for the same reason. Do you use a lot of different cables? Then colored cables may be helpful, using different colors for different lengths or types of cable. You can also color code cables yourself.

Plugs

If a cable gets noisy or dies all of a sudden, it's usually a conductor that has come loose within one of the plugs. Soldering plugs requires practice, the right tools, a steady hand, and knowing how the conductors should be wired. Unsure? Check the wiring on the plug on the other end of the cable. Incorrect wiring may cause hum, noise, or total silence. When you turn a balanced cable into an unbalanced one, you may even damage your equipment (see page 164). *Tip:* there are solderless ¼" phono plugs!

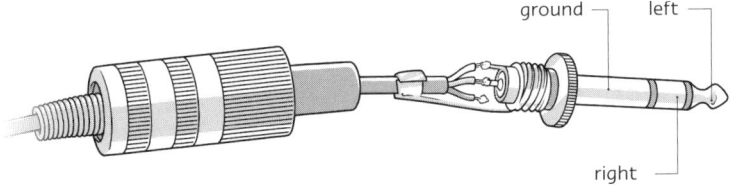

ground — left —

right —

Wiring of a TRS phone plug (top) and a balanced XLR plug.

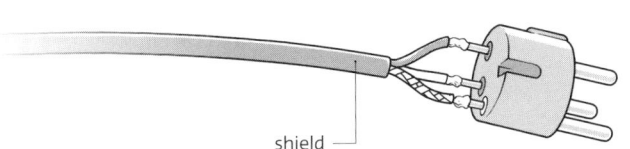

shield —

245

CLEANING

Keeping the outside of your equipment clean doesn't take much more than a soft, lint-free cloth. You can moisten it a little with water, or use a mild, non-abrasive household cleaner. Always spray cleaners onto a cloth, instead of directly onto the unit: The cleaner should not get inside the unit or its controls. Never use solvents. Carpet-covered enclosures can be cleaned with a stiff brush and/or a vacuum cleaner.

Sockets
Keeping your plugs clean helps keep your jacks clean. Noisy contacts? Moisten a cotton swab with some rubbing alcohol and clean the inside of the jack.

Noisy pots
Noisy pots can be cured with contact spray, but this may require opening up the unit. Just spraying a lot of contact lubricant in the general direction of the pot can do more harm than good.

Microphone grilles
Vocal mics can get quite dirty. Cleaning them requires removing and replacing the grille and the built-in pop filter, which may not be easy. When in doubt, contact a technician or ask your dealer for advice.

HUM AND NOISE

Hum and noise can be the result of poor wiring, neon lights, dimmers, refrigerators and other devices with an electrical engine, or nearby radio stations or power transformers, just to name a few.

- Noise can be prevented by keeping signal cables (instrument cables, etc.) away from devices that generate **electromagnetic interference** (EMI), such as televisions and monitors, adapters, and power wiring — so never tie power cables and signal cables together.

• The effects unit, the instrument, or the amp you're using may also be the culprit. Disconnect all devices and connect them **one at a time** to see where things go wrong. Keep volume controls down when (dis)connecting things.

• Cables often induce hum or noise. Try **replacing long cables with short ones** (especially when they're unbalanced), and remember that good cables can be worth the extra money in this respect too.

• Use **DI boxes** to turn unbalanced signals into balanced signals, and hi-Z signals into lo-Z signals, and use the ground lift switch to prevent ground hum. Any other solution is a dangerous one, as stated on page 61.

• Some amps have a **polarity switch**, which turns the AC's + into − and vice versa. This may reduce hum. In some countries, you can simply invert the power plug for the same effect.

• Powerline noise can be solved with **a power conditioner** (see page 74).

• Coiled up speaker cables may pick up nearby **radio stations**. Try coiling them up the other way around.

Feedback

Feedback is a type of noise that has been dealt with in previous chapters. Here's a brief compilation:

• If built-in notch filters and phase switches don't help, you may check out **dedicated anti-feedback units.**

• Selecting the right microphones and experimenting with **mic placement** (close miking, or adjusting mic angles) can be effective.

• You may solve feedback problems by using **in-ear monitors.**

• **Equalizers** can reduce feedback, but if the range they affect is too large (more than ⅓ octave; see page 46), your sound will suffer.

247

- Acoustic guitarists may try a special **feedback buster**, i.e., a rubber disc that blocks the instrument's soundhole.

- If nothing else seems to work, all you can do is **lower the overall or stage volume levels**.

TUBE AMPS

Tube amps require special care, as their tubes need to be replaced from time to time. Depending on how often and how loud you play, and on how carefully you treat your amp, power tubes usually last between one and three years. Preamp tubes typically last twice as long.

When?

Signal loss, a reduced treble response and an unclear bass, undesirable distortion, and (microphonic) noise tell you that it's time to replace your tubes. Tube prices range from about ten to forty dollars.

Your technician

You can replace tubes yourself, but it's usually safer to have it done. Why? Because expensive and potentially dangerous mistakes are easily made, especially when it comes to power tubes. Amps use lethal voltages.

Bias

Tubes need a basic voltage to keep them 'idling.' This voltage is adjusted with the amp's bias setting. If the *bias* is set too low, tubes will get even hotter than they're supposed to (regular internal temperatures run up to some 700°F/370°C!) As a result, the sound will lack punch and hum may occur. Disagreeable distortion and lack of power? Then the bias setting may be too high.

Auto-bias, fixed bias, and matching tubes

Some amps adjust bias settings automatically, allowing you to experiment with different types of tubes for different sounds.

248

Even so, you'll generally be advised to consult your technician. *Fixed-bias* amps need tubes that match the old ones, as their bias can't be adapted to other tubes. Power tubes are usually supplied in *matched sets* of two, four, six, or eight, with each tube in a set requiring the same bias.

ON THE ROAD

Taking your equipment on the road? Here are some tips:

Front or back
Place combos and speaker cabinets on their front or back, so bumps in the road make speaker cones move the way they're supposed to. Note that the equipment should not rest on its controls, of course. Also make sure that equipment can't fall over or slide, damaging controls and other vulnerable parts.

Amp covers
The number of amps that come with a matching, padded cover is quite limited. Covers offer some protection, but if you travel often, a hard-shell case is a better investment. These *flight* or *road cases* are expensive and heavy, but very effective.

Accessories
Mics, cables, tuners, effects units, receivers, transmitters, and other accessories are best stored in padded pouches or cases, which may be included. The better protected they are, the longer they will last, and the higher the price you can fetch if you want to sell them.

Spare items
These are the main spare items you should bring along:

- An extra **cable** for any type of cable you need, including power cables.

- Spare **fuses** of the correct ratings.

- One or more **extension cords** with a multi-outlet power strip.

- **Batteries** for effects units, transmitters and receivers, and so on.

- Spare **adapters**.

- … and **everything** you need to keep playing your instrument.

Carrying stuff around

You can save yourself a lot of trouble and pain by using a handtruck for your heavier devices. They don't cost a lot, and they're available in basic, folding, and convertible models, with either pneumatic (shock-absorbing, but not puncture-proof) or solid wheels. Some amps and cabinets come with casters, but these aren't usually designed for outside use.

Insurance

Consider insuring your equipment, especially if you're taking it on the road. Musical instruments and everything else covered in this book usually fall under the 'valuables' insurance category. A regular homeowner insurance policy will not cover all possible damage, whether it occurs at home, on the road, in the studio, or onstage.

17

History

In the past, instruments were made to sound louder by simply making them bigger, or by using more of them: A symphony orchestra sounds as loud as it does because there are so many musicians. And then came the amplifier.

The very first tubes were made in the early 1900s. Initially, they were used for radios. Funnily enough, the first loudspeaker was patented in the 1870s, before there were amplifiers. The first 'modern' version of the loudspeaker came around fifty years later.

Pioneer

Guitar amplification presumably dates back to the 1930s, Charlie Christian being one of its early proponents. The first guitar amps had an 8" or 10" speaker, a very modest power rating, a volume control, a tone control, and that was it.

Fifties and sixties

It wasn't until the 1960s that guitar amps were provided with equalizers and a spring reverb. The integrated tremolo was introduced around 1955. In that same era, Celestion designed the first dedicated guitar speaker, replacing the radio speakers that had been used so far. That same speaker, the Alnico Blue, is still being made today.

Classic designs

This is not the only design that has withstood the test of time. Two other examples are the Fender Bassman, a 1959 40 watt model with four 10" speaker, developed for bassists but especially popular among guitarists; or the Vox AC30 with 36 watts or power and two 12" speakers, from that same period.

More power

Around 1965, the Who's Pete Townshend asked Marshall to build him a 100 watt amp — unheard of in those days — with eight 12" speakers. The combined size and weight of the speakers inspired another debut, the stack: two cabinets and a head. A couple of years later Ampeg introduced their legendary 300 watt SVT tube bass amp.

Solid state

The first solid-state amps were presented in the early 1960s. They weren't very successful because of both their sound and reliability. It wasn't until the early 1980s that solid-state amps seemed to take over from tube amps — but tube amps never disappeared and they probably never will. One of the few classic solid-state amps is the 1975 2x60 watt Roland Jazz Chorus.

More firsts

In 1972, Mesa made the first successful guitar amp with a separate master volume control, allowing for heavy distortion at lower volume levels. The first two-channel amps with channel switching were presumably made around that time too. The first headphone guitar amp, the walkman-sized SR&D Rockman, was introduced in the early 1980s.

Modeling

Roland introduced modeling in 1993 with their VG8. Line 6 debuted their digital modeling amplifier three years later.

Acoustic amps and keyboard amps

The first keyboard amps were pioneered by Peavey and Roland around 1980. Trace Elliot claims to have made the first acoustic amp some ten years later.

Sound systems

Over the years, sound systems have become increasingly powerful. In the mid-1960s, anything over a couple of hundred watts was a lot, and the musical performance of these systems left a lot to be desired. Today, sound systems with a power rating of 200,000 watts and more are not uncommon.

Effects

Fuzz boxes and wah-wahs were some of the first guitar effects, besides the built-in spring reverbs and tremolos. Electro-Harmonix was one of the first dedicated effect makers. Another major name is Boss, closely related to Roland, who made their first phasers, distortion, and wah pedals in the late 1970s. Around that same time, Lexicon made the first digital reverb, at a price that only professional studios could afford. Multi-effects became popular in the 1980s.

18

Brands

There are numerous companies that make amps, effects, and related products. Some are highly specialized, while other companies make musical instruments too. This chapter introduces you to some of the names you will come across.

Many of the brand names and companies listed below will be around for years to come — but some may be discontinued by the time you read this, while others may have changed ownership and/or product lines, or the price range they focus on.

In other words, this is the most time sensitive chapter of this book, and it's not intended to be complete. Please refer to musician's magazines and websites (see pages 274–276) for up-to-date information.

Fender and Marshall

When it comes to amplifiers, the two main names are **Fender** and **Marshall**. Both companies make amps in a wide range of prices. Marshall specializes in guitar amps and effects, while Fender produces various types of amps as well as their well-known guitars and basses. Fender is as well known for 'the' American guitar sound as Marshall is for 'the' British sound. Leo Fender, who never learned to play an instrument, built his first guitars and amps in his radio shop, back in the early 1940s. Drummer (!) Jim Marshall, known as 'the father of loud,' introduced his first guitar amp some twenty years later.

Peavey

Peavey, founded by Hartley Peavey in 1965, is one of the other big names in the world of amps. The company makes a large variety of amplifiers in almost all price ranges, as well as sound systems and microphones, electric and acoustic guitars, and bass guitars.

Low to medium budget

A lot of companies focus on low- to medium-budget amps, say up to some five or six hundred dollars for a guitar combo. **Behringer** (Germany) also makes studio equipment, sound and lighting systems, microphones, and a host of related products; **Alesis** and **Zoom** are well known for their effects; **Epiphone** (owned by Gibson), **Dean**, **Ibanez**, **Samick**, and **Washburn** are renowned guitar and bass guitar makers. Some other names are **Aria**, **Dinosaur**, **Drive**, **Park** (owned by Marshall), **Sovtek** (from Russia), **Stagg**, and **Traynor** (from the Canadian company that also makes Bassmaster amps, and Yorkville sound systems and keyboard amps).

Up to a thousand and more

Other companies have guitar combos from about a hundred or less to a thousand or more dollars, such as Crate (makes sound systems too), **Kustom**, **Laney** (UK), and **Randall. Tech21**, **Rocktron**, and **Line6** start a little higher. British amp makers **Vox** and **Hiwatt** (UK) build much more expensive amps too. **Carvin**, **Hughes & Kettner** (the largest German amp maker), **Gibson** (guitars, of course), **Orange**, and **Ashdown** (started with bass amps only) start between some four and six hundred dollars, going up to two thousand dollars and, in some cases, even more. **Carlsbro**, from England, also makes PA systems. **Polytone**, especially popular among jazz players, makes guitar and bass amps in a limited price range only.

Higher

And of course, there are companies that mainly produce amps in the higher price range, their lowest priced combo amps easily costing a thousand dollars or more. **Mesa Engineering** (Mesa/Boogie; 1970) is one of the largest companies in this category. Some other names are **Bogner**, **Budda**, **Brunetti** (Italy), **Egnater**, **Evans**, **Groove Tubes**, **Holland**, **Rivera**, **Soldano**, **THD Electronics**, and the German companies **Diezel**, **Engl**, and **Framus. Koch** (co-designs for Eden and Sadowsky) and **Marble** are made in the Netherlands.

Amps too

Two of the largest companies in the music industry make amps as a relatively small part of their business: **Yamaha** is one, and **Roland** is the other. Roland is best known for their keyboards and other electronic instruments. Yamaha makes numerous instruments — from horns to pianos and guitars — as well as motorcycles, sailboats, and much, much more.

Bass amps and more

Many of the companies listed above make both guitar and bass amps, and some also make other types of amps as well as monitors and related products. Many companies focus primarily on bass amps. Most of them start in the medium price range, while some offer lower priced amps as well. A few examples are **Aguilar**, EBS

257

(Swedish), **Eden** and **Nemesis** (both from the same company), **Euphonic Audio**, **Audiophile** and **Mark Bass** (both from Italy), **SWR** (makes acoustic amps too), **Glockenklang** (German; also makes sound systems), **Hevos** (a Dutch company), **Gallien-Krueger, Phil Jones Pure Sound** (USA), and **Walter Woods** (specializes in upright bass amps). **Ampeg**, **Genz Benz**, **Hartke** and **Trace Elliott** also make guitar amps. **Warwick** produces both bass amps and bass guitars.

Acoustic amps
Acoustic amps are built by a number of companies mentioned above (Ashdown and Marshall, for example) and by a limited number of dedicated companies such as Fishman (well-known for their transducers), **High Cliff** (replacing the speaker with a wooden soundboard), and **AER** (makes guitar and bass amps too). Trace Elliot and Carlsbro were two of the acoustic amp pioneers.

Keyboard amps
Keyboard amps are made by some of the companies listed earlier in this chapter, such as Roland, Behringer, Carlsbro, Crate, Peavey, and Yorkville, but also by **Barbetta** (makes PA and DJ systems too) and specialized makers such as **Motion Sound**.

Sound systems
When looking for sound systems, you'll come across some of the names above, of course. A few other players in this field are **Electro-Voice**, **HK Audio**, **LEM**, **Solton**, and **Dynacord**. **Soundcraft** is well known for their mixers.

Effects
You don't need a large factory to produce stompboxes (or amps, for that matter) which helps explain why there are so many brand names in the world of effects. **Boss** is one of the main names, having sold millions of compact effects since 1976. Quite a few amp makers make effects too, such as **Fender**, **Ibanez**, and **Vox**. A few more names you're bound to come across are **Morley** (mainly volume and wah-pedals), **Carl Martin** (Danish), **Danelectro**, **DOD**, **Electro-Harmonix** (one of the first specialists), **Roger Mayer** (UK), **Korg** (well known for their keyboard instruments),

258

and **Jim Dunlop**. **Digitech** makes multi-effects units as well as digital and analog stompboxes, while **Zoom** focuses on multi-effects and a host of other equipment. **Rocktron** and **TC Electronic** are best known for their rack effects. **Alesis** makes guitar and audio processors as well as mixers, drum machines, synths, sound modules, and other products.

Speakers

Most amplifier makers buy their speakers from specialized companies such as **Celestion** and **Fane** (UK), **Electro-Voice**, **Eminence**, **JBL**, and **Jensen**. **Electro-Voice** makes microphones too.

Microphones

Well-known microphone brands include **AKG**, **Apex**, **Audio-Technica**, **Audix**, **Beyerdynamic**, **Crown**, **Electro-Voice** (**EV**), **Røde**, **Samson**, **Sennheiser**, and **Shure**, as well as some of the names mentioned earlier. **Neumann**, **Brauner**, **sE Electronics**, **Studio Projects**, and **MXL** focus on studio microphones.

Glossary

This glossary briefly explains all the guitar jargon used in this Tipbook, and also contains some terms that haven't been mentioned, but which you may come across in other books, magazines, or online. Most terms are explained in more detail as they are introduced in this book. Please consult the index on pages 280–284.

¼" phone plug, ¼" plug
See: *Phone plug.*

1x15, 2x10, 4x12
Speaker cabinets with one 15" speaker, two 10" speakers, etc. Also indicated as 115, 210, etc.

19" equipment
See: *Rack.*

A/B box
Small device with one input, two outputs, and a switch to select the A or B output.

Acoustic amplifier
Amplifier for acoustic instruments, e.g., acoustic guitars, vocals, horns, etc.

Active EQ
See: *EQ, equalizer.*

Active speaker
Speaker with built-in power amplifier(s).

Adapter
Mainly used in this book to indicate a device that converts AC voltage from a power outlet to DC voltage suitable for effects and other units. Also known as *power supply unit* (*PSU*).

A/D converter
Converts analog signals to bits.

Algorithm
The 'formula' or 'program' for a digital effect. A device with ten reverb algorithms provides ten different digital reverbs (e.g., spring 1, spring 2, room, or church).

All-tube amp
See: *Tube amp.*

Ambient effects
Reverb and delay are ambient effects.

Amplifier
An amplifier consists of two main sections. The *preamplifier* or *preamp* is used to shape and control the sound by operating its volume, tone, and other controls. It's also known as the *control amp*. The *power amp* amplifies the signal from the preamp and sends it to the speakers. An *integrated amp* has the preamp and power amp in one housing; separate preamps and power amps are also available.

Audio level
See: *Sound pressure level.*

Aux, auxiliary
1. Extra input.
2. Connections and controls for effects and/or monitors, used on mixers.

Backline
Stage equipment: the musicians' own amps and effects, which are not part of the sound system.

Baffle board
The board onto which the speakers are mounted.

Balanced connection
Noise suppressing connection; also known as a symmetrical connection.

Bandwidth
See: *EQ, equalizer.*

Bell-shape EQ
See: *Shelving EQ.*

Bi-amping
Using separate amps for the bass and treble ranges.

Bias
A basic — adjustable — voltage that keep amp tubes 'idling.'

262

Bit depth
Resolution. A larger bit depth (i.e., more bits, or a higher resolution) allows for finer detail.

Bridging
In the bridged mode, a 2x100 watt stereo amp behaves like a 200 watt mono amp.

Buffered
A buffered effects loop prevents impedance mismatches. Buffered inputs that can handle various signal and impedance levels.

Cab, cabinet
Enclosure with one or more speakers.

Cannon plug
See: *XLR plug.*

Cardioid pattern
See: P*ickup pattern.*

Channels
Mixers and various types of amps have two or more channels, allowing for the amplification of multiple instruments, or (on guitar amps) for multiple sounds. Stereo instruments require two channels to sound in stereo.

Chorus
Effect.

Class A, Class A/B
A class A/B amps, pairs of tubes work in an alternating *push-pull* manner, dividing the workload, so to speak. In a *class A* amp, all of the tubes continuously run at full power.

Class H
A weight-saving amplifier design with reduced power consumption. Powered speakers often use class H amps.

Clipping
If you overdrive an amp, it will 'clip' the sound. A clip indicator shows you when clipping occurs.

Closed back
Combos have either a closed or an open back.

Coax(ial) speaker
See: *Dual cone speaker.*

Coaxial cable
Instrument cables are unbalanced, coaxial cables with a shielded center conductor.

Combo, combo amp
Amplifier and speaker(s) in one housing.

Compressor
Effect.

Concentric speaker
See: *Dual cone speaker.*

Cone
The part of a speaker that moves the air.

Contour
Additional tone-shaping control. Also known as *shape* or *voicing.*

Control amp
See: *Preamp.*

Crossover filter
Sends the low frequencies to the woofer, and the high frequencies to the tweeter (and the midrange frequencies to a midrange driver, if there is one).

D/A converter
Converts bits to analog signals.

dB, decibel
Unit used to measure sound pressure and various other levels.

263

De-esser
Effect that reduces the sizzling effects of the vowel 's' and other sibilant sounds

Depth
Effect parameter.

DI
Low impedance, balanced output to connect an amp or effects unit directly to a sound system or a recording console. See also: *DI box*.

DI box
Direct-injection box or *direct box*. Turns a high-impedance, unbalanced signal into low-impedance, balanced signal.

Direct box, direct injection
See: *DI box*.

Distortion
An overdriven amp generates a distorted sound. Unwanted by most other musicians, but a desired effect for many guitarists, as well as some bassists and other musicians. See also: *Overdrive (OD)*.

Driver
Speaker.

Dry
See: *Wet*.

DSP
Digital Signal Processor.

Dual concentric speaker
See: *Dual cone speaker*.

Dual cone speaker
Space-saving design; the high frequency driver is centered in the low frequency driver.

Dynamic effects
Effects that affect volume.

Dynamics
The difference between loud and soft.

Effects loop
Lets you connect effects between the preamp and the power amp. Loops can be serial (the entire signal goes through the loop) or parallel (there are two signal paths: one goes through the loop, the other stays in the amp).

EMI
Electromagnetic interference, generated by devices such as monitors and TV sets.

Enclosure
See: *Cabinet*.

EQ, equalizer
Tone control. Most amps have a three-band equalizer, allowing you to cut (and boost, if it's an *active EQ*) three bands or frequency ranges: bass, midrange, and treble. With a *semi-parametric*, *quasi-parametric* or *sweepable EQ*, you can choose the frequency range you want to affect. A parametric EQ also lets you set the (*band*) *width*, *slope* or Q of that range. A *rotary* EQ has *rotary controls*; a *graphic* EQ uses faders, graphically showing the overall EQ setting, as illustrated on page 45.

Expander
Effect.

Expression pedal
An expression pedal can be used to control volume level, chorus depth, wah-wah and other continuous parameters. Also known as *modulation pedal* or *rocker pedal*.

Fader
Sliding control.

Feedback
1. Loud squealing or howling sound,

caused by instruments or microphones that pick up the sound from a speaker, send that back to amp, pick up the boosted sound again, and so on.
2. Effect parameter.

Flanger
Effect.

FOH
See: *Front-of-house.*

Foldback
See: *Monitor speakers.*

Foot controllers
Foot controllers can be used to change channels or presets, switch effects on and off, boost volume, and so on.

Frequency response
A microphone with a linear frequency response picks up all frequencies equally well, producing an uncolored, faithful reproduction of the original sound. A shaped frequency response boosts some frequencies and cuts others. Speakers and other devices have either a linear or a shaped frequency response too.

Front-of-house
The front-of-house speakers are aimed at the audience.

Full stack
See: *Stack.*

FX
Effects.

Gain
Controls the signal level that is fed to the preamp. Can be used to generate distortion (which is what guitarists often do), or to prevent it (what others usually do), matching the instrument's output level to the amp. Also known as *trim control.*

Graphic equalizer
See: *Equalizer.*

Ground lift switch
The only safe way to cure ground hum.

Grounding
Equipment that comes with a three-prong power plug needs to be grounded by using a properly wired power outlet.

Half stack
See: *Stack.*

Harmonics
Each note you play consists of a fundamental frequency and a mixture of harmonics, the latter determining the character or timbre of the sound.

Headroom
The 'reserve' of an amplifier.

Hertz (Hz)
Unit of frequency. At 440Hz, a string vibrates 440 times per second.

Hex pickup, hexaphonic pickup
Pickup with a separate output for each string, used in combination with guitar synthesizers.

Hi-trim
Treble control.

Horn
A horn provides tweeters with a wider sound dispersion.

HT fuse
Heat fuse.

Hybrid amps
Amps that use both solid-state and tube technology.

Hypercardioid pattern
See: *Pickup pattern.*

265

Hz
See: *Hertz.*

Impedance (Z)
(Electric) instruments, speakers, microphones, pickups, and other devices all have a certain impedance or resistance, sometimes indicated as Z, and measured in ohms (Ω). Impedance mismatches may result in a degraded signal, noise, or damage.

In-ear monitors
Can help prevent feedback and hearing damage.

Integrated amplifier
Most amps are integrated, having both the preamp and the power amp in one housing.

Jack
See: *Socket.*

k, kHz
1k or kHz equals 1,000Hz.
See: *Hertz.*

Latching switch
Switch that is used to turn things on or off. A non-latching switch is like a door buzzer, turning the power on as long as you hold it.

Leslie speaker
Speaker cabinet with rotating horn and drum.

Limiter
Prevents loud signals from causing distortion or damaging the speaker.

Line in
Input for line-level signals.

Line level signals
There are two standard line levels: –10dB (consumer electronics, keyboard instruments, etc. Also known as *instrument level*) and +4dB (rack effects and other pro-audio equipment).

Line out
Output; connects to the line in of an amp or mixer.

Loop
See: *Effects loop.*

Lo-trim
Bass control.

Mid, midrange
The frequency range between bass and treble.

MIDI
Short for Musical Instrument Digital Interface. MIDI-equipped digital musical instruments and other devices (amps, effects units, computers) can be hooked op to one another so they can 'communicate.'

Mixer
Device to 'mix' the volume and other levels of different sound sources, creating a balanced overall sound.

Modeling
Technique to (digitally) emulate the characteristics of amplifiers, speaker cabinets, effects, microphones, or instruments.

Modulation effects
Effect category, including chorus, flanger and other *time-based* effects as well as tremolo and vibrato, for example.

Modulation pedal
See: *Expression pedal.*

Monitor speakers
The speakers that address the band, as

opposed to the *front-of-house* speakers. Originally known as *foldback (speakers)*.

Multi-effects unit
Unit with multiple, programmable effects.

Music power
See: *Watts RMS.*

Mute switch
Turns the sound off.

Noise gate
Effect.

Non-latching switch
See: *Latching switch.*

Notch filter
Suppresses feedback by filtering out the offending frequency.

Octaver
Effect.

OD
See: *Overdrive.*

Ohm
See: *Impedance.*

Omnidirectional
See: *Pickup pattern.*

Open back
See: *Closed back.*

Output
An output connects to a matching input.

Overdrive (OD)
Many guitarists, and some bassists, keyboardists and other musicians overdrive their amp to add a little or a lot of fuzz or distortion to the sound. An overdrive effect produces a similar result.

PA
Public Address. See: *Sound system.*

Pad
Attenuator. Can be used to prevent input overload or to lower a tweeter's volume level, for example.

Pan, panorama
Controls the placement of an instrument or a voice in a stereo image.

Parallel
Speakers , effects , and other devices can be wired either in parallel or in series.

Parallel loop
See: *Effects loop.*

Parameter
Effects can be controlled by adjusting their parameters.

Parametric equalizer
See: *EQ, equalizer.*

Passive EQ
See: *EQ, equalizer.*

Passive speaker
See: *Powered.*

Patch
See: *Preset.*

Peak power
See: *Watts RMS.*

Peaking EQ
See: *Shelving EQ.*

Phantom power
'Invisibly' powering microphones and other devices by using the cable that also transports the audio signal.

Phase (inversion) switch
Switch that helps fight feedback, or

267

brings signals back in phase so they don't cancel each other out.

Phaser
Effect.

Phone plug
Plug with a ¼" pin, commonly used on instrument cables; also incorrectly known as *phono plug*.

Phono plug
Another name for RCA plug. See also: *Phone plug*.

Pickup
Small device, attached to or built into your instrument; picks up what you play and transmits those signals to an amp. Pickups are also known as transducers.

Pickup pattern
Microphones are available with various pickup patterns. The pattern indicates the angle(s) at which a mic picks up, reduces, or rejects sound. There are unidirectional (cardioid, supercardioid, hypercardioid), omnidirectional, and other types of microphones.

Pitch correction
Corrects pitch as you play or sing.

Pitch shifter
Effect.

Plate reverb
Effect.

Ports
Cabs and (keyboard and bass) combos often have one or more ports that enhance the unit's (bass) performance.

Pot, Potentiometer
When you turn a volume or tone knob,

you operate the pot or potentiometer on the other side of the panel. The pot is the actual device that lowers or raises the volume and other settings.

Power amplifier
See: *Amplifier*.

Power attenuator
Soaks up all or part of the power of a tube amp, allowing you to overdrive the power section at low volume levels (or to use a headphone on a tube amp).

Power conditioner
Protects equipment and fights line noise.

Power rating
An amp's power or a speaker's maximum power handling , expressed in watts.

Powered
Powered mixers and speakers have a built-in power amp. Powered speakers are also known as *active speakers*, as opposed to ampless *passive speakers*.

Preamp, preamplifier
See: *Amplifier*.

Predelay
Effect parameter. Postpones the effect.

Presence
Control to influence tone. Described in may different ways.

Preset
On digital devices, you can usually store your settings in a number of presets or *patches*.

Proximity effect
Due to this effect, getting closer to a microphone generates extra bass and warmth.

PSU
See: *Adapter.*

Q
See: *EQ, equalizer.*

Quasi-parametric equalizer
See: *EQ, equalizer.*

Rack
A rack is a frame that holds equipment (amps, effect units, mixers, and so on) with a standard width of 19".

Rate
Effect parameter, adjusting the 'speed' of an effect.

RCA
RCA plugs and jacks are commonly used on (home) stereo systems, as well as for auxiliary connections.

Receiver
See: *Wireless systems.*

Receptacle
See: *Socket.*

Reverb, reverberation
Effect.

RFI
Radio frequency interference.

Ring modulator
Effect.

RMS
See: *Watts RMS.*

Rotary equalizer
See: *Equalizer.*

Rotor
Effect.

Semi-parametric equalizer
See: *EQ, equalizer.*

Serial loop
See: *Effects loop.*

Series
See: *Parallel.*

Shape
See: *Contour.*

Shelving EQ
Influences the entire range above or below a certain frequency, as opposed to a bell or peaking EQ that influences a certain frequency and a given range around it.

Sidefill
Monitor speaker on the side of the stage.

Signal path
The route from one device to the other. Should be kept as short as possible.

Signal-to-noise ratio
Indicates the ratio between the noise and the signal produced by amps and other devices.

Slew rate
An amp with a high slew rate responds very quickly to its input, generating a dynamic, tight sound with lots of nuance.

Slope
See: *EQ, equalizer.*

Socket
You plug your plugs into sockets, jacks, or receptacles.

Solid state
Solid-state equipment uses solid-state (ss) or 'silicon' technology (transistors or semiconductors), rather than tubes.

Sound pressure level (SPL)
Audio level. How loud things sound;

269

expressed in *decibels*. Also used to indicate how much sound a microphone can handle or how much sound a speaker can produce .

Sound system
Basically a mixer, an amplifier and two or more speakers. Also known as a PA. Larger sound systems also include monitor speakers for the band, a separate monitor mixer, and more.

Spatial sound
Effect.

Speaker emulation, speaker simulation
Emulates the effect of a (guitar) speaker cabinet when connecting an amp or an effects unit directly to a mixing or recording console, or when using headphones.

Speakon
High-quality speaker connector.

SPL
See: *Sound pressure level.*

Spring reverb
Effect, commonly found in guitar amps.

Ss
See: *Solid state.*

Stack
A stack is made up of one (half stack) or two (full stack) speaker cabinets and a head.

Stage
High-gain tube amps can have up four or more gain stages.

Stompbox
Compact effects unit with footswitch(es).

Subwoofer
Speaker that reproduces the very

lowest (e.g., bass and bass drum) frequencies.

Supercardioid pattern
See: *Pickup pattern.*

Sustainer
Effect.

Sweepable EQ
See: *EQ, equalizer.*

Symmetrical
See: *Balanced.*

Tone controls
See: *EQ, equalizer.*

Transducer
See: *Pickup.*

Transient
Brief, loud (attack) sound.

Transmitter
See: *Wireless systems.*

Treble
The high frequency range.

Threshold
Effect parameter.

Trim
See: *Gain*

TRS plug
Phone plug with three conductors; tip, ring, and sleeve. Allows for stereo and balanced connections, as opposed to a (mono) TS plug.

True bypass
Basically a straight wire that runs though the unit, allowing the signal to go directly from the input jack to the output so it cannot be degraded by the circuitry.

TS plug
See: *TRS plug.*

Tube
Amps, microphones and other devices that use vacuum tubes or *valves* typically produce a warmer, more natural sound than solid-state equipment. Tube guitar amps are popular for their warm, smooth distortion.

Tube amp
Amplifier with tubes both in the preamp and power amp sections. Sometimes referred to as *all-tube amp*, as opposed to (hybrid) amps that have tubes in either the preamp or the power amp section.

Tweeter
Dedicated driver for high frequencies. Often combined with a *horn.*

Twin cone speaker
See: *Dual cone speaker.*

UHF
Frequency range for signal transmission of wireless systems.

Unbalanced
See: *Balanced connection.*

Unidirectional
See: *Pickup pattern.*

Vacuum tube
See: *Tube.*

Valve
See: *Tube.*

VHF
Frequency range for signal transmission of wireless systems.

Voice coil
Part of the speaker that moves the cone.

Wah-wah
Effect.

Watt
Unit of power.

Watts RMS
The *continuous average power* of an amp is expressed in watts RMS (Root Mean Square). Music power and peak power ratings of that same amp are much higher. The power handling of a speaker is commonly expressed in watts RMS too.

Wedge
Monitor speakers are often wedge shaped.

Wet
Parallel effects loops and certain effect devices let you balance the wet (effected) and dry (uneffected) signal.

Whammy
Effect.

Wireless systems
A wireless system uses a transmitter and a receiver instead of a microphone or instrument cable.

Woofer
Dedicated speaker for the low frequency range.

XLR plug
Round plug with three pins, commonly used for balanced connections. Also known as Cannon plug.

Z
See: *Impedance.*

271

Tipcode List

The Tipcodes in this book offer easy access to short videos, sound files, and other additional information at www.tipbook.com. For your convenience, the Tipcodes in this Tipbook have been listed below.

Tipcode	Topic	Page	Chapter
Amps-001	Distortion	8	2
Amps-002	Reverb, chorus, wah-wah, flanger, delay	23	2
Amps-003	Various amp models	25, 103	2, 7
Amps-004	Various equalizer settings guitar amp	44, 111	4, 7
Amps-005	Various equalizer settings (piano)	44	4
Amps-006	Shape settings bass amp	50	4
Amps-007	Various guitar speakers	73, 98	4, 7
Amps-008	A 3dB difference	81	5
Amps-009	The range of a piano	89	5
Amps-010	British, American, and modern	99	7
Amps-011	Presence	112	7
Amps-012	EQ settings and your bass sound	131	8
Amps-013	Various types of reverb	174	12
Amps-014	Various types of delay	176	12
Amps-015	Chorus, phaser, flanger	180	12
Amps-016	Overdrive	182	12
Amps-017	Distortion	182	12
Amps-018	Wah-wah	186	12
Amps-019	Auto-wah	187	12
Amps-020	Pitch shifter and octaver	188	12
Amps-021	Effects chain 1	197	12
Amps-022	Effects chain 2	197	12

Want To Know More?

Of course, there's a lot more to read on all the subjects on the previous pages, in magazines, in books, and online. If you want to know more about your instrument or other instruments, then consult the other volumes in the Tipbook Series.

MAGAZINES

Guitar and bass magazines offer reviews of amps, effects, and related gear, and the same goes for dedicated keyboard and drum magazines. Names and contact information can be found in the relevant volumes of the Tipbook Series (see pages 285–289), or visit www.musicianmagazines.com for examples of US magazines. Other publications of interest include magazines on pro-audio (sound systems, etc.) and recording, such as:

- *AudioMedia* (www.audiomedia.com)

- *Electronic Musician* (www.emusician.com)

- *EQ* (www.eqmag.com)

- *Future Music* (UK; www.futuremusic.co.uk)

- *Mix* (www.mixonline.com)

- *Music Tech Magazine* (UK; www.musictechmag.co.uk)

- *Pro Audio Review* (www.proaudioreview.com)

- *Professional Sound* (Canada; www.professional-sound.com)

- *Prosound News* (www.prosoundnews.com)

- *Recording* (www.recordingmag.com)

- *Sound on Sound* (UK; www.soundonsound.com)

BOOKS

Most books on amps and effects are on guitar amps and effects only — and there are dozens of them around. The number of books that deal with other types of amps is extremely limited, but there are many publications on live sound and sound systems. Here are some examples, as well as some books on microphones. Please note that this list may not include the latest additions in this field, and that some of the books listed here may have been updated when you read this.

- *The Acoustic Musician's Guide to Sound Reinforcement and Live Recordings*, Mike Sokol (Prentice Hall, 1997; 288 pages; ISBN 978-0134335094).

- *Basic Live Sound*, Paul White (Sanctuary Press, 2004; 208 pages; ISBN 978-1860742712).

- *The Basics of Live Sound: Tips, Techniques & Lucky Guesses*,   Jerry J. Slone (Hal Leonard, 2002; 120 pages; ISBN 978-0634030284).

- *Live Sound: PA for Performing Musicians*, Peter Buick (Cimino Publishing Group, 1997; 200 pages; ISBN 978-1870775441).

- *Live Sound for Musicians*, Rudy Trubitt (Hal Leonard, 1997; 152 pages; ISBN 978-0793568529).

- *The Live Sound Manual: Getting Great Sound at Every Gig*, Ben Duncan (Backbeat Books, 2002; 160 pages; ISBN 978-0879306991).

- *Live Sound Reinforcement; A Comprehensive Guide to PA and Music Reinforcement Systems and Technology*, Scott Hunter Stark (Artistpro, 2004; 380 pages; ISBN 978-1592006915).

- *Sound Reinforcement Handbook*, Gary D. Davis and Ralph Jones (Hal Leonard, 1990; 418 pages; ISBN 978-0881889000).

- *Basic Microphones*, Paul White (Sanctuary Press, 2004; 224 pages; ISBN 978-1860742651).

- *Basic Mixers*, Paul White (Sanctuary Press, 2004; 192 pages; ISBN 978-1860742668).

- *The Microphone Book*, John Eargle (Focal Press, 2001; 368 pages; ISBN 978-024080445).

- *Professional Microphone Techniques*, David Mills Huber and Philip Williams (Artistpro, 1999; 152 pages; ISBN 978-0872886858).

Guitarists

Guitarists have an almost unlimited choice of books on amps and effects. There are books on legendary amp makers; on repairing, servicing, and modifying (tube) amps; on vintage amps and effects; and even on instrument, amp, and effect settings. Searching your book store for 'guitar amp', 'guitar sounds' or 'guitar effects' will provide you with numerous titles.

275

INTERNET

There's a wealth of information to be found online, as long as you know where to start. If you want to learn what others think of the amplifier that you're about to buy, you may simply search for the name of the amp ('brand name model') and the word 'review'. As with books, most websites on amps are on guitar amps. Here are some starting points. Do note that some of the websites of the magazines listed earlier have a lot to offer too!

- www.amptone.com

- www.geofex.com

- www.guitargeek.com

- www.guitarnoise.com

- www.harmonicalessons.com/mics.html

- www.harmony-central.com

- www.harpamps.com

- www.musiciansnews.com

- www.songstuff.com

- www.soundonsound.com

- www.tubefreak.com

- www.ultimate-guitar.com

- www.wholenote.com

Glossaries

The Internet offers lots of glossaries as well, for example at www.audioed.com.au, www.themusicedge.com, and www.soundonsound.com. Glossaries can also be found on the websites of manufacturers, distributors, music stores, etc.

Essential Data

In the event of your equipment being stolen or lost, or if you decide to sell it, it's useful to have all the relevant data at hand. Here are two pages to list everything you need – for insurance purposes, for the police, for a prospective buyer, or just for yourself.

INSURANCE

Company: _____ Phone: _____

Broker: _____ Phone: _____

Email: _____ Website: _____

Policy number: _____ Premium: _____

Renewal date: _____

AMPS, EFFECTS UNITS, AND OTHER EQUIPMENT

Make and model: _____

Serial number: _____

Price: _____ Date of purchase: _____

Dealer: _____ Phone: _____

Email: _____ Website: _____

Make and model: _____

Serial number: _____

Price: _____ Date of purchase: _____

Dealer: _____ Phone: _____

Email: _____ Website: _____

Make and model: _____

Serial number: _____

Price: _____ Date of purchase: _____

Dealer: _____ Phone: _____

Email: _____ Website: _____

277

Make and model:

Serial number:

Price: Date of purchase:

Dealer: Phone:

Email: Website:

Make and model:

Serial number:

Price: Date of purchase:

Dealer: Phone:

Email: Website:

Make and model:

Serial number:

Price: Date of purchase:

Dealer: Phone:

Email: Website:

ADDITIONAL NOTES

ESSENTIAL DATA

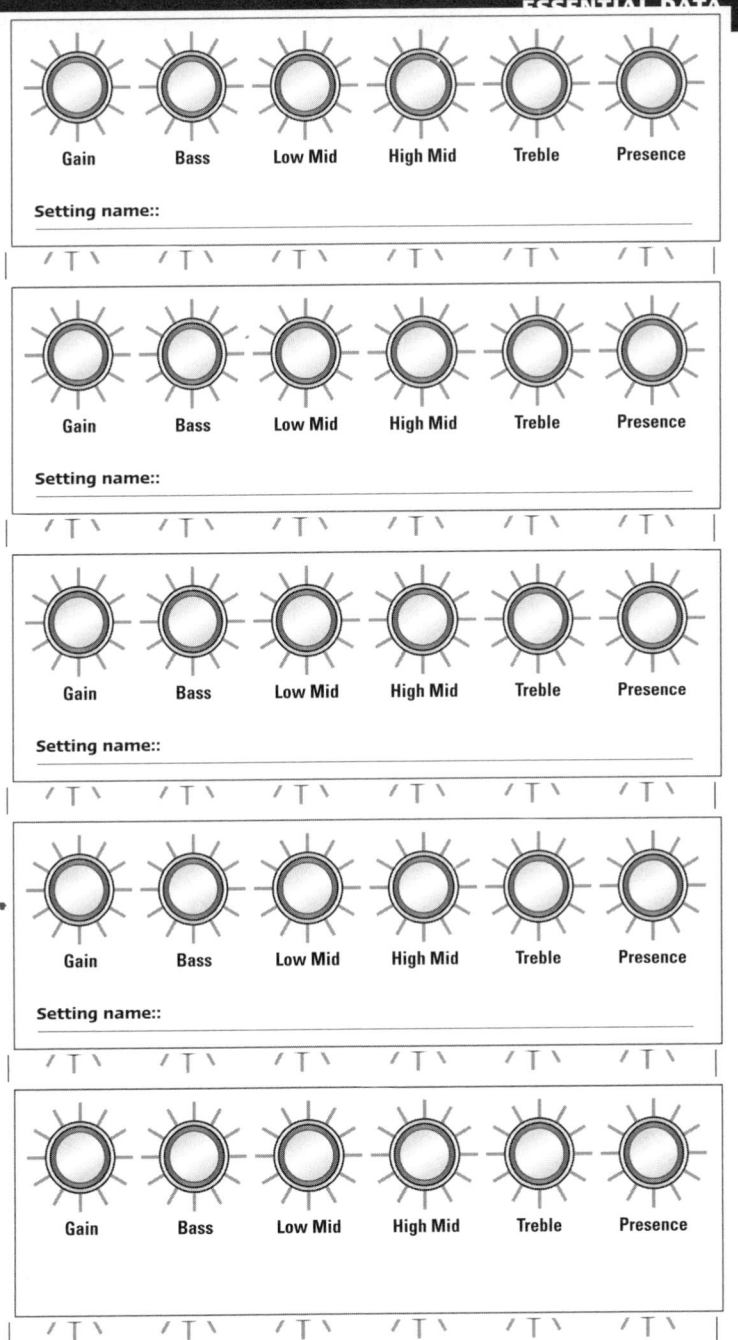

Gain   Bass   Low Mid   High Mid   Treble   Presence

Setting name::

Gain   Bass   Low Mid   High Mid   Treble   Presence

Setting name::

Gain   Bass   Low Mid   High Mid   Treble   Presence

Setting name::

Gain   Bass   Low Mid   High Mid   Treble   Presence

Setting name::

Gain   Bass   Low Mid   High Mid   Treble   Presence

Gain   Bass   Low Mid   High Mid   Treble   Presence

Setting name::

279

SETTINGS

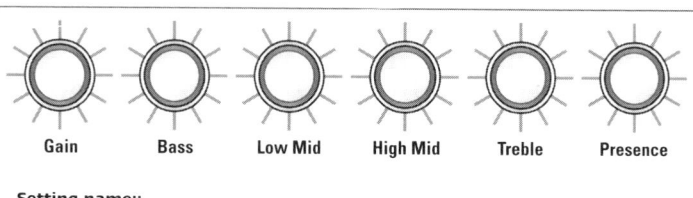

Gain   Bass   Low Mid   High Mid   Treble   Presence

Setting name:: _____

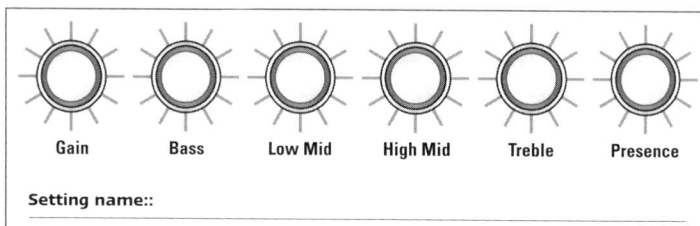

Gain   Bass   Low Mid   High Mid   Treble   Presence

Setting name:: _____

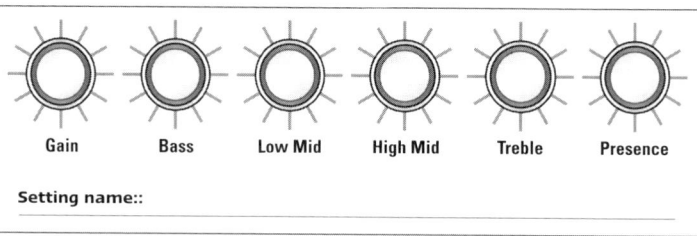

Gain   Bass   Low Mid   High Mid   Treble   Presence

Setting name:: _____

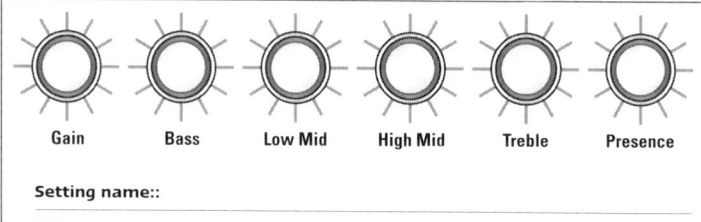

Gain   Bass   Low Mid   High Mid   Treble   Presence

Setting name:: _____

Gain   Bass   Low Mid   High Mid   Treble   Presence

Setting name:: _____

SETTINGS

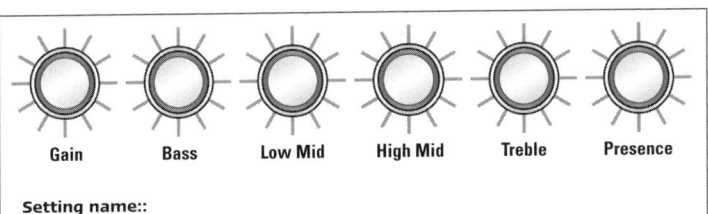

Gain   Bass   Low Mid   High Mid   Treble   Presence

Setting name::

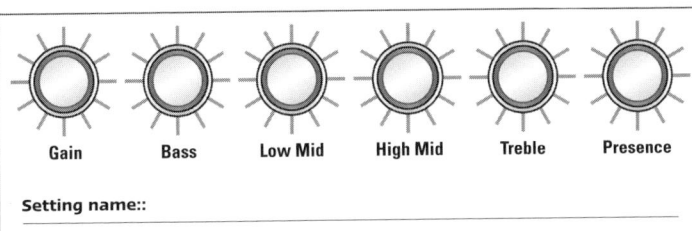

Gain   Bass   Low Mid   High Mid   Treble   Presence

Setting name::

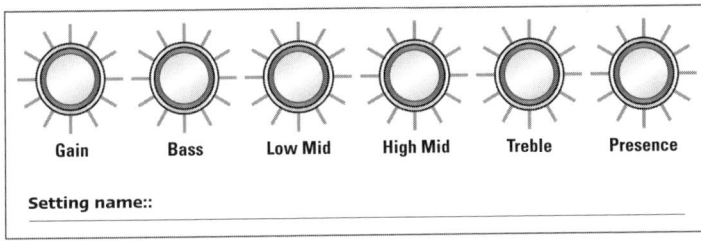

Gain   Bass   Low Mid   High Mid   Treble   Presence

Setting name::

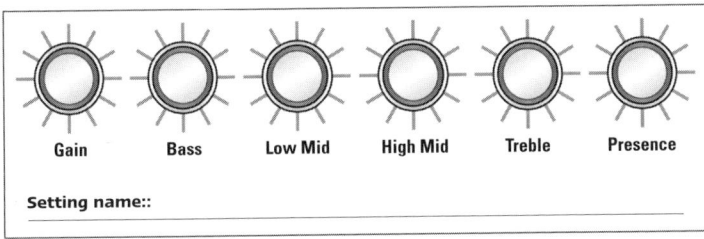

Gain   Bass   Low Mid   High Mid   Treble   Presence

Setting name::

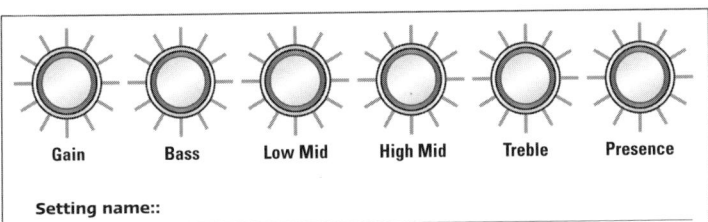

Gain   Bass   Low Mid   High Mid   Treble   Presence

Setting name::

Index

Please check out the glossary on pages 261–271 for additional definitions of the terms used in this book.

¹/₄" phone plug, ¹/₄" plug: 51–52, 54–55, 87, 221, 230, 245
19" equipment: 15, 24–25, 65
1x15, 2x10, 4x12: 14

A

A/B box: 109
A/D converter: 68, 262
AC: 39–40, 242–243
Acoustic amplifier: 19, 29–30, 65, 88, 137–144
Active bass, guitar: 126–127
Active DI box: 60, 61
Active EQ: 48–49
Active input: 109
Active speaker: 152
Adapter (plug): 52, 54
Adapter (power supply unit): 60, 203, 210
AES/EBU: 69
AFL: 158
Algorithm: 280
Alnico: 72
Ambience, Ambient effects: 172, 174–175
American sound: 98
Amplitude modulation: 190
Analog: 67, 68, 186, 195
Audio interface: 16
Audio level: 81, 237
Aux control: 156, 160–161
Aux, auxiliary: 9, 55, 60, 141, 150, 160–161

B

Backline: 154, 262
Backplate: 216
Baffle board: 11
Balance control: 41, 157, 190, 223
Balanced connection: 53–54, 60, 88, 221, 230, 235, 245
Bandwidth: 47, 142
Bargraph meters: 159
Bass roll-off switch: 221
Batteries: 13, 60, 127, 202–203, 212
Bell-shape EQ: 48
Bi-amping: 79, 128, 166–167
Bias: 248–249
Bit depth: 69
Booster: 182
Bridging: 79, 153
British sound: 98, 99
Buffered: 65, 141, 263
Burst power: 78

C

Cab emulation: 57
Cab, cabinet: 9, 10–11, 13,–14, 25, 28, 36, 38, 80, 84–85, 105, 114–115, 128–130
Cable ears: 39
Cable length: 54, 196, 231, 232, 235, 245
Cables: 16, 43, 51, 54, 55, 58, 230–236
Cannon: 53
Capacitance: 232
Capsule: 216
Cardioid pattern: 218–219
Cartridge: 216

Casters: 38

Center-negative polarity: 210

Ceramic magnets: 72

Channel fader: 156, 157, 161

Channel gain: 156

Channels: 11–12, 17, 20, 21, 107–108, 126, 140–141, 148–149, 155–158

Chicken-head pointer knobs: 41

Chorus: 23, 119, 143, 172, 179, 196–198

Class A, Class A/B: 117

Class D: 67, 126

Class H: 167

Clip light: 50, 58–59

Clipping: 58, 100–101, 116

Closed back: 105, 114–115

Coax(ial) speaker: 73

Coaxial cable: 232

Combo, combo amp: 2, 7, 10, 13–14

Compressor: 132, 190–192, 193, 196–197

Compressor/sustainer: 191

Concentric speaker: 73

Condenser microphone: 164, 216–217, 223, 226

Condenser pickups: 226

Cone tweeters: 73

Cone: 70–72, 73, 125, 129

Continuous controller: 189, 204

Contour: 51

Courtesy outlet: 40

Crossover (filter): 133, 167, 263

Crunch: 17

D

D/A converter: 68

Damping factor: 102

dB: 48, 81–84, 86–87, 113

DC: 103, 164, 203, 210

Decibel: 48, 81–84, 86–87

De-esser: 201

Delay effects: 178

Delayed attack: 190

Depth: 173

Detuner: 189

DI box: 54, 60–61, 88, 141, 164

DI: 59–62

Diaphragm: 216

Digital equipment: 67–70

Digital output: 211

Digital signal processor (DSP): 68

Direct box: 60

Direct injection: 59–62

Direct out: 59

Directivity response: 218

Distortion: 8, 17, 58, 87, 89, 99–100, 102, 107, 115–117, 125, 126, 129, 157, 159, 181–184

Distortion pedal: 107, 139

Diversity: 240

Driver: 10

Dry: 63, 65, 173, 174, 194

DSP: 68

Dual concentric speaker: 73

Dual cone speaker: 73, 140

Ducking: 177–178

Dynamic effects: 172, 190

Dynamic microphone: 164, 216–217

Dynamic tweeters: 73

Dynamics: 63, 74, 88, 92, 101, 186, 190–192, 197

E

Early reflections: 174

Earth lift switch: 61

Earthing: 242

Effect looper: 198

Effects loop: 62–65, 133, 195–198

Efficiency: 83, 126

EMI: 232

Envelope filter, envelope follower: 187

Envelope: 187

EQ, equalizer: 6, 42, 43–51, 61–62, 93, 99–100, 110–112, 131–132, 142, 158, 159, 184

283

EQ panel diagrams: *49, 206*
Expander: *148, 193*
Expression pedal: *25, 185, 201, 204*
Extension speaker: *9, 55, 80, 84–86*

F

Fader throw: *157*
Fader: *21, 155, 156, 157*
Fans: *74*
Feedback: *19, 87, 142–143, 144,*
   168–169, 174, 219, 244, 247
Ferrite: *72*
Fixed-bias amps: *249*
Flanger: *179–180*
FOH: *21, 162, 169*
Foldback: *161*
Foot controllers: *16, 25, 42–43*
Frequency response: *113, 220*
Front-of-house: *21, 162, 165*
Full stack: *13–14, 114*
Full-range fader: *157*
Fuzz box: *23, 139, 181, 197, 203, 253*
FX: *62*

G

Gain: *7–8, 17, 21, 58–59, 99, 108–110,*
   111, 117, 120, 156–157, 182, 191–192
Gauge (cables): *234, 235*
GND (Ground): *62*
Graphic equalizer: *44, 45–46*
Ground lifter, ground lift switch: *61,*
   167, 247
Ground lift: *61*
Ground hum: *60, 247*
Grounding: *40, 61, 242, 247*

H

Half stack: *13, 14*
Handles: *37*
Handling power: *80–81*

Hard clipping: *100, 116*
Hard distortion: *115*
Hard knee: *191*
Harmonica amplifiers: *18*
Harmonics: *89–90, 101, 131, 172,*
   180, 239
Headphone jack: *7, 8–9, 51, 52, 56, 60,*
   117, 143, 213
Headphones: *57*
Headroom: *79, 103, 116, 117, 127–128,*
   133
Heat fuse: *74*
Hertz (Hz): *89*
Hex pickup, hexaphonic pickup: *213*
High-end pad: *140*
Hi-trim: *6*
Horn: *10, 73*
HT fuse: *74*
Hum: *51, 55, 58, 60, 61, 88, 134, 152,*
   167. 227, 242, 245, 246–248
Hybrid amps: *66, 103, 125*
Hypercardioid pattern: *218*
Hz: *89*

I

Impedance (Z): *57, 65, 84–86, 87–88,*
   101–102, 113, 133–134, 141, 153, 221
In-ear monitors: *169*
Inputs: *8–10*
Insert: *9, 63, 163*
Integrated amplifier: *11*

J

Jack: *9, 10*

K

k, kHz: *89*

L

Latching switch: 43
Late reflections: 174
LED ladder: 157, 159
Leslie speaker: 149, 181
Limiter: 132, 192
Line in, Line input: 86
Line level (signals): 65, 86–87, 235
Line out: 9, 58, 59–60
Linearity: 124, 132, 220
Long-excursion speakers: 72
Loop switcher: 198
Loop: 62–65, 133, 195–198
Lo-trim: 6
Low-cut switch: 221

M

Master effect: 160
Master section: 21, 154, 155, 159–160
Master volume control: 6
Microphone modeling: 25, 144
Microphone preamp: 200, 217–218
Microphones: 59, 87–88, 216–225
MIDI: 69–70, 105, 200, 211
Mini-jack plug: 52
Mixer: 20–22, 152–153, 154–164
Mixing loop: 63
Modeling: 25, 66, 103–106, 108, 120, 124, 143–144, 172, 201, 202, 208–209, 253
Modern sound: 98
Modulation effects: 178, 181, 190, 196
Monitor sends: 161
Monitor speakers: 22, 38, 160, 161–163, 165, 168
Multi-effects unit: 24, 30, 36, 160, 204–206, 206–208
Music power: 78
Mute switch: 58

N

Nearfield speakers / monitors: 147, 164

Neodymium: 72
Noise gate: 167, 193, 194
Noise: 50, 58, 60, 63, 69, 74, 86, 94–95, 121, 134, 152, 221, 225, 230, 231, 232, 233, 236, 239, 242, 246–248
Nominal level: 157
Non-diversity: 240
Non-latching switch: 43
Notch filter: 19, 60, 135, 141, 142

O

Octaver: 132, 188
Ohm: 84
Omnidirectional: 133, 218
Open back: 105, 114–115
Output: 8–10
Overdrive (OD): 17, 89, 98, 100–101, 108–109, 115, 118, 125, 172, 181, 182, 184, 196–197

P

PA: 3, 20–22, 59, 83, 139, 150, 151–170
Pad: 61, 140
Pan, panorama: 156, 157
Panner: 190
Parallel loop: 63–64, 65
Parallel (speakers): 85–86
Parallel (effects) 197–198
Parameter: 23, 24, 25
Parametric equalizer: 47–48, 131, 185
Passive bass: 126, 135
Passive DI box: 60
Passive EQ: 48–49
Passive input: 109
Passive speaker: 152, 161
Patch: 24, 68
Peak power: 78
Peaking EQ: 48
Pedal board: 70, 203
Pentode: 116
PFL: 156, 158

Phantom power: 60, 127, 141, 164, 217–218, 225–226

Phase (inversion) switch: 19, 135, 141, 142–143

Phaser: 178–179, 180

Phone plug: 55

Phono plug: 55

Phrase trainer: 13, 212

Pickup pattern: 218–219

Pickup: 87, 88, 105, 133–134, 141, 213, 225–227

Piezo tweeters: 73

Piggyback: 14

Pingpong effect: 176, 190

Pitch correction: 189, 201

Pitch shifter: 23, 172, 188–189

Plate reverb: 174, 175

Point-to-point wiring: 75

Polar pattern: 218

Polarity:210

Ported design, ports: 129–130

Post gain: 110

Post-fader listening: 156, 158

Pot, Potentiometer: 41, 156, 246

Power amplifier: 11, 12, 21, 64, 115–116

Power attenuator: 116

Power conditioner: 74

Power handling: 80–81

Power rating: 78–80

Powered mixer: 21, 152, 161,

Powered speakers: 21, 135, 152, 166

Pre/post EQ switch: 62

Preamp, preamplifier: 11, 17, 21, 25, 58, 62, 64, 95, 103, 133–134, 141, 172, 183, 208–209, 217–218

Predelay: 173

Pre-fader listening: 156–158

Presence: 6, 111

Preset: 167–168, 194, 200, 206–208

Proximity effect: 221

PSU: 210

Public Address: 3

Push-pull: 117

Q

Q: 47–48

Quasi-parametric equalizer: 48

R

Rack wings: 155

Rack, Rack-mounted equipment:15, 24–25, 51, 155, 200, 201, 209–210

Rate: 173

RCA plugs, sockets: 52, 55, 162–163

Real time: 25, 189, 204

Receiver: 236, 237–238, 239, 240

Receptacle: 10

Recording out: 60

Rectifier: 103

Release: 191, 192

Resolution: 69

Resonance control: 102, 174, 180

Response pattern: 218

Return (from FX): 62

Reverb tail: 175

Reverb, reverberation: 23, 110, 119, 160, 174–175, 197, 201, 252

RFI: 232

Ribbon microphones: 217

Ring modulator: 181

RMS: 12, 78

Rotary equalizer:44

Rotor: 150, 181

RU: 15

S

Sampling rate: 68

Semi-parametric equalizer: 47, 48

Send (to FX): 62

Serial loop: 63–64, 163

Series, in – (speakers): 80, 85–86

Series, in – (effects): 197–198

Shape: 51

Shelving EQ: 48

Side chain loop: 63

Signal path: 54, 57, 134, 196

Signal-to-noise ratio: 86

Skirted knobs: 41

Slanted cab: 114

Slew rate: 269

Slope: 47

Slot tweeters: 73

Socket: 10

Soft clipping: 100, 116, 118, 125

Soft distortion: 115

Soft knee: 191

Software: 15–16, 66, 67, 69, 70, 105

Solid state amps: 65-66, 84, 89,
   100–103, 106–108, 124–125, 133, 139,
   182, 252–253

Solo switch: 43

Sound dispersion pattern: 73, 93

Sound pressure level (SPL): 83

Sound systems: 3, 20–22, 59, 83, 139,
   150, 151–170

Spatial sound: 150, 181

SPDIF: 69

Speaker emulation, speaker

simulation: 57–58, 60, 102, 103, 209

Speakers: 2, 10–11, 12, 70–73, 80–81,
   84–86, 95, 112–114, 128–129, 140,
   147, 164–169

Speakon: 55

Spillover: 177

SPL: 83, 113

Spring reverb: 18, 119, 252

Squelch: 239

Ss: 65

Stack: 13–14, 114

Standing waves: 95

Stereo: 8, 13, 20, 29, 51–52, 65, 79, 114,
   139, 147–148, 150, 155, 156–157, 194

Stompbox: 24, 30, 62, 65, 200,
   201–204, 204–206, 210, 237

Stop points: 208

Stove knobs: 41

Subwoofer: 11, 22, 72, 147, 165–167

Supercardioid pattern: 218, 219

Sustainer: 191

Sweepable EQ, Swept EQ: 48, 142,
   158, 185

Symmetrical: 53

T

Tank: 119

Tap tempo: 42, 127

Thermal protection circuit: 74

Threshold: 173, 187, 191, 192, 193

Time-based effects: 178

Tone controls: 6, 12

Tracking: 189

Transducer: 270

Transient: 270

Transmitter: 236–239

Triboelectric impact noise: 233, 236

Trim control: 6, 21, 156

Triode: 116

Triode/pentode switch: 116

TRS (plug): 52

True bypass:

True diversity: 240

TS (plug): 52, 54, 163, 245

Tube: 118–119, 182–183, 201

Tube amp: 66, 85, 100–103, 106–107,
   115–119, 124–125, 243, 248–249

Tube preamp: 148, 182–182, 208

Tweeter: 10–11, 19, 73, 74, 79, 93, 129,
   140, 166

Twin cone speaker: 73

U

U: 15

UHF: 239

Unbalanced: 54, 60, 221, 232, 245, 247

Unidirectional: 218, 222

USB: 16, 200, 211

287

V

Vacuum tube: 66
Valve: 37, 66
VHF: 239
Voice coil: 70, 71

W

Wah-wah: 63, 172, 185–187, 204, 253
Watt: 12–13, 78–79, 80–81, 82, 83, 84,
    106, 116, 127, 139, 147, 152, 153, 154,
    166–167
Watts RMS: 12, 78
Wedge: 38
Wet: 63, 65, 173, 174, 194
Whammy: 189

Wireless systems: 236–240
Woofer: 10, 73, 79, 133, 147, 165
Workstation: 200

X

XLR (plug, connections): 52, 53, 54–55,
    87, 245

Y

Y-cable: 163

Z

Z: 8 7, 141, 232, 247

The Tipbook Series

*Did you like this Tipbook? There are also Tipbooks for your fellow
band or orchestra members! The Tipbook Series features various
books on musical instruments, including the singing voice, in
addition to Tipbook Music on Paper, Tipbook Amplifiers and
Effects, and Tipbook Music for Kids and Teens – a Guide for
Parents.*

*Every Tipbook is a highly accessible and easy-to-read
compilation of the knowledge and expertise of numerous
musicians, teachers, technicians, and other experts,
written for musicians of all ages, at all levels, and in any
style of music. Please check www.tipbook.com for up to
date information on the Tipbook Series!*

*All Tipbooks come with Tipcodes that offer additional
information, sound files and short movies at www.tipbook.com*

Instrument Tipbooks

All instrument Tipbooks offer a wealth of highly accessible, yet well-founded
information on one or more closely related instruments. The first chapters of
each Tipbook explain the very basics of the instrument(s), explaining all the
parts and what they do, describing what's involved in learning to play, and
indicating typical instrument prices. The core chapters, addressing advanced
players as well, turn you into an instant expert on the instrument. This
knowledge allows you to make an informed purchase and get the most out of
your instrument. Comprehensive chapters on maintenance, intonation, and
tuning are also included, as well a brief section on the history, the family, and
the production of the instrument.

Tipbook Acoustic Guitar – $14.95

Tipbook Acoustic Guitar explains all of the elements
that allow you to recognize and judge a guitar's
timbre, performance, and playability, focusing on
both steel-string and nylon-string instruments.
There are chapters covering the various types of
strings and their characteristics, and there's plenty
of helpful information on changing and cleaning
strings, on tuning and maintenance, and even on the
care of your fingernails.

289

Tipbook Amplifiers and Effects – $14.99

Whether you need a guitar amp, a sound system, a multi-effects unit for a bass guitar, or a keyboard amplifier, *Tipbook Amplifiers and Effects* helps you to make a good choice. Two chapters explain general features (controls, equalizers, speakers, MIDI, etc.) and figures (watts, ohms, impedance, etc.), and further chapters cover the specifics of guitar amps, bass amps, keyboard amps, acoustic amps, and sound systems. Effects and effect units are dealt with in detail, and there are also chapters on microphones and pickups, and cables and wireless systems.

Tipbook Cello – $14.95

Cellists can find everything they need to know about their instrument in *Tipbook Cello*. The book gives you tips on how to select an instrument and choose a bow, tells you all about the various types of strings and rosins, and gives you helpful tips on the maintenance and tuning of your instrument. Basic information on electric cellos is included as well!

Tipbook Clarinet – $14.99

Tipbook Clarinet sheds light on every element of this fascinating instrument. The knowledge presented in this guide makes trying out and selecting a clarinet much easier, and it turns you into an instant expert on offset and in-line trill keys, rounded or French-style keys, and all other aspects of the instrument. Special chapters are devoted to reeds (selecting, testing, and adjusting reeds), mouthpieces and ligatures, and maintenance.

Tipbook Electric Guitar and Bass Guitar – $14.95

Electric guitars and bass guitars come in many shapes and sizes. *Tipbook Electric Guitar and Bass Guitar* explains all of their features and characteristics, from neck profiles, frets, and types of wood to different types of pickups, tuning machines, and — of course — strings. Tuning and advanced do-it-yourself intonation techniques are included.

Tipbook Drums – $14.95

A drum is a drum is a drum? Not true — and *Tipbook Drums* tells you all the ins and outs of their differences, from the type of wood to the dimensions of the shell, the shape of the bearing edge, and the drum's hardware. Special chapters discuss selecting drum sticks, drum heads, and cymbals. Tuning and muffling, two techniques a drummer must master to make the instrument sound as good as it can, are covered in detail, providing step-by-step instructions.

Tipbook Flute and Piccolo – $14.99

Flute prices range from a few hundred to fifty thousand dollars and more. *Tipbook Flute and Piccolo* tells you how workmanship, materials, and other elements make for different instruments with vastly different prices, and teaches you how to find the instrument that best suits your or your child's needs. Open-hole or closed-hole keys, a B-foot or a C-foot, split-E or donut, inline or offset G? You'll be able to answer all these questions — and more — after reading this guide.

Tipbook Keyboard and Digital Piano – $14.99

Buying a home keyboard or a digital piano may find you confronted with numerous unfamiliar terms. *Tipbook Keyboard and Digital Piano* explains all of them in a very easy-to-read fashion — from hammer action and non-weighted keys to MIDI, layers and splits, arpeggiators and sequencers, expression pedals and multi-switches, and more, including special chapters on how to judge the instrument's sound, accompaniment systems, and the various types of connections these instruments offer.

Tipbook Music for Kids and Teens – a Guide for Parents – $14.99

How do you inspire children to play music? How do you inspire them to practice? What can you do to help them select an instrument, to reduce stage fright, or to practice effectively? What can you do to make practice fun? How do you reduce sound levels and

291

prevent hearing damage? These and many more questions are dealt with in *Tipbook Music for Kids and Teens – a Guide for Parents and Caregivers.* The book addresses all subjects related to the musical education of children from pre-birth to pre-adulthood.

Tipbook Music on Paper – $14.99

Tipbook Music on Paper – Basic Theory offers everything you need to read and understand the language of music. The book presumes no prior understanding of theory and begins with the basics, explaining standard notation, but moves on to advanced topics such as odd time signatures and transposing music in a fashion that makes things really easy to understand.

Tipbook Piano – $14.99

Choosing a piano becomes a lot easier with the knowledge provided in *Tipbook Piano*, which makes for a better understanding of this complex, expensive instrument without going into too much detail. How to judge and compare piano keyboards and pedals, the influence of the instrument's dimensions, different types of cabinets, how to judge an instrument's timbre, the difference between laminated and solid wood soundboards, accessories, hybrid and digital pianos, and why tuning and regulation are so important: Everything is covered in this handy guide.

Tipbook Saxophone – $14.95

At first glance, all alto saxophones look alike. And all tenor saxophones do too — yet they all play and sound different from each other. *Tipbook Saxophone* discusses the instrument in detail, explaining the key system and the use of additional keys, the different types of pads, corks, and springs, mouthpieces and how they influence timbre and playability, reeds (and how to select and adjust them) and much more. Fingering charts are also included!

Tipbook Trumpet and Trombone, Flugelhorn and Cornet – $14.99

The Tipbook on brass instruments focuses on the smaller horns listed in the title. It explains all of the jargon you come across when you're out to buy or rent an instrument, from bell material to the shape of the bore, the leadpipe, valves and valve slides, and all other elements of the horn. Mouthpieces, a crucial choice for the sound and playability of all brasswinds, are covered in a separate chapter.

Tipbook Violin and Viola – $14.95

Tipbook Violin and Viola covers a wide range of subjects, ranging from an explanation of different types of tuning pegs, fine tuners, and tailpieces, to how body dimensions and the bridge may influence the instrument's timbre. Tips on trying out instruments and bows are included. Special chapters are devoted to the characteristics of different types of strings, bows, and rosins, allowing you to get the most out of your instrument.

Tipbook Vocals – The Singing Voice – $14.95

Tipbook Vocals –The Singing Voice helps you realize the full potential of your singing voice. The book, written in close collaboration with classical and non-classical singers and teachers, allows you to discover the world's most personal and precious instrument without reminding you of anatomy class. Topics include breathing and breath support, singing loudly without hurting your voice, singing in tune, the timbre of your voice, articulation, registers and ranges, memorizing lyrics, and more. The main purpose of the chapter on voice care is to prevent problems.

International Editions

The Tipbook Series is also available in Spanish, French, German, Dutch, Italian, and Chinese. For more information, please visit us at www. tipbook.com.

Tipbook Series Music and Musical Instruments

Tipbook Acoustic Guitar
ISBN 978-1-4234-4275-2, HL00332802 — $14.95

Tipbook Amplifiers and Effects
ISBN 978-1-4234-6277-4, HL00332776 — $14.99

Tipbook Cello
ISBN 978-1-4234-5623-0, HL00331904 — $14.95

Tipbook Clarinet
ISBN 978-1-4234-6524-9, HL00332803 — $14.99

Tipbook Drums
ISBN 978-90-8767-102-0, HL00331474 — $14.95

Tipbook Electric Guitar and Bass Guitar
ISBN 978-1-4234-4274-5, HL00332372 — $14.95

Tipbook Flute & Piccolo
ISBN 978-1-4234-6525-6, HL00332804 — $14.99

Tipbook Home Keyboard and Digital Piano
ISBN 978-1-4234-4277-6, HL00332375 — $14.99

Tipbook Music for Kids and Teens
ISBN 978-1-4234-6526-3, HL00332805 — $14.99

Tipbook Music on Paper — Basic Theory
ISBN 978-1-4234-6529-4, HL00332807 — $14.99

Tipbook Piano
ISBN 978-1-4234-6278-1, HL00332777 — $14.99

Tipbook Saxophone
ISBN 978-90-8767-101-3, HL00331475 — $14.95

Tipbook Trumpet and Trombone, Flugelhorn and Cornet
ISBN 978-1-4234-6527-0, HL00332806 — $14.99

Tipbook Violin and Viola
ISBN 978-1-4234-4276-9, HL00332374 — $14.95

Tipbook Vocals — The Singing Voice
ISBN 978-1-4234-5622-3, HL00331949 — $14.95

Check www.tipbook.com for additional information!